YANMAR MARINE DIESEL ENGINE
1SM/2SM/3SM

SERVICE MANUAL

YANMAR MARINE DIESEL ENGINE 1SM/2SM/3SM

SERVICE MANUAL

ISBN/EAN: 9783954275090
Erscheinungsjahr: 2012
Erscheinungsort: Bremen, Deutschland

© maritimepress in Europäischer Hochschulverlag GmbH & Co. KG, Fahrenheitstr. 1, 28359 Bremen. Alle Rechte beim Verlag und bei den jeweiligen Lizenzgebern.

www.maritimepress.de | office@maritimepress.de

Bei diesem Titel handelt es sich um den Nachdruck eines historischen, lange vergriffenen Buches. Da elektronische Druckvorlagen für diese Titel nicht existieren, musste auf alte Vorlagen zurückgegriffen werden. Hieraus zwangsläufig resultierende Qualitätsverluste bitten wir zu entschuldigen.

YANMAR
SERVICE MANUAL

MODEL
1SM/2SM/3SM

CONTENTS

PREFACE 1
1. OUTLINE OF STRUCTURES 2
2. FUEL AND LUBRICATING OIL 3
   2-1. Fuel 3
   2-1-1. Light Oil 3
   2-1-2. Properties of Fuel & Engine Performances 5
   2-1-3. Cautions on Fuel 6
   2-2. Lubricating Oil 6
   2-2-1. Engine Oil 6
3. SERVICING AND PERIODICAL CHECKING 8
   3-1. Main Points of Servicing and Training 8
   3-2. Time of Periodical Checking 11
4. ENGINE DISASSEMBLY 13
   4-1. Preparation for Disassembly 13
   4-2. Precautions for Disassembly 16
   4-3. Disassembling Procedure 16
5. ENGINE REASSEMBLY 30
   5-1. Precautions for Reassembly 30
   5-2. Reassembling Procedure 30
6. DISASSEMBLY AND REASSEMBLY OF OTHER PARTS 48
   6-1. Fuel Injection Pump 48
   6-2. Fuel Injection Valve 53
   6-3. Cylinder Head 56
   6-4. Replacement of Cylinder Liner 58
   6-5. Piston Rings, Piston and Connecting Rod 60
   6-6. Clutch Housing 64
7. ADJUSTMENT AND SERVICING 67
   7-1. How to Adjust the Adjusting Part 67
   7-2. Servicing and Repairing of Major Parts 75
   7-3. Maintenance Standards of Main Parts 80
8. COUNTERMEASURES TO ENGINE TROUBLES 86
   8-1. Engine Troubles, Causes and Measures 86
   8-2. Causes and Measures for Reduction
        Reversing Gear 94
9. STORAGE OF ENGINE 95
10. LIST OF APPROVED OILS 96

EXTERNAL DIMENSIONS (1SM)

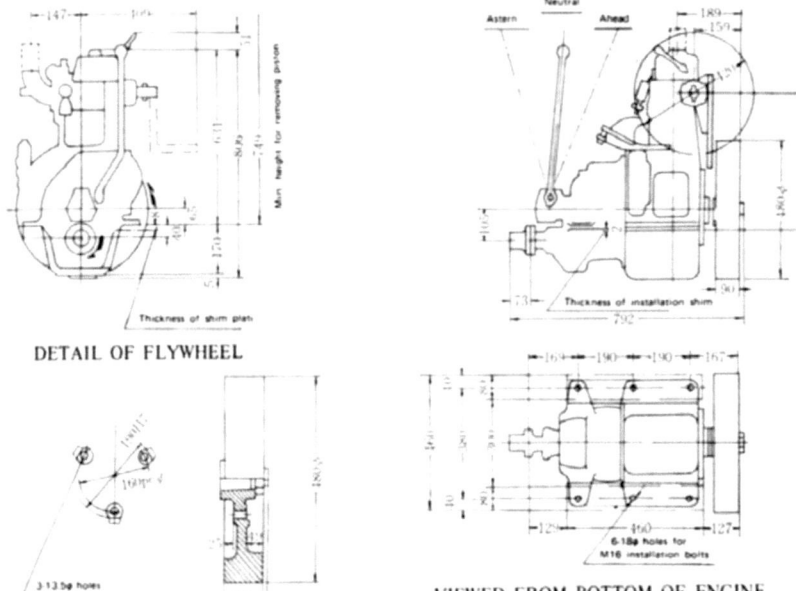

DETAIL OF FLYWHEEL

VIEWED FROM BOTTOM OF ENGINE

PARTS LOCATION (1SM)

EXTERNAL DIMENSIONS (2SM)

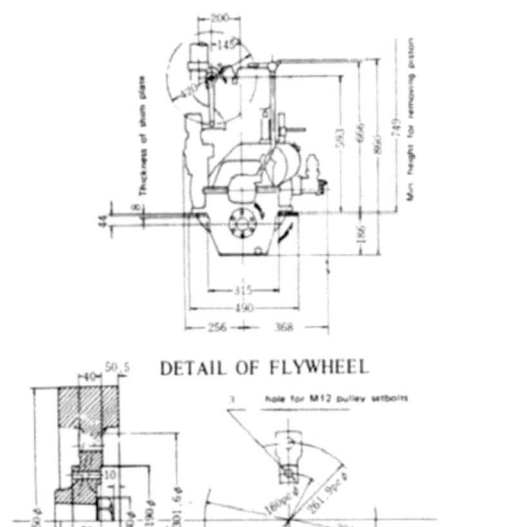

DETAIL OF FLYWHEEL

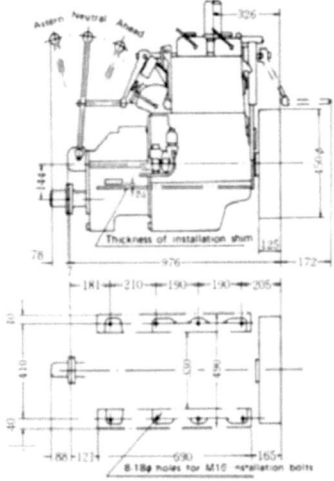

VIEWED FROM BOTTOM OF ENGINE

PARTS LOCATION (2SM)

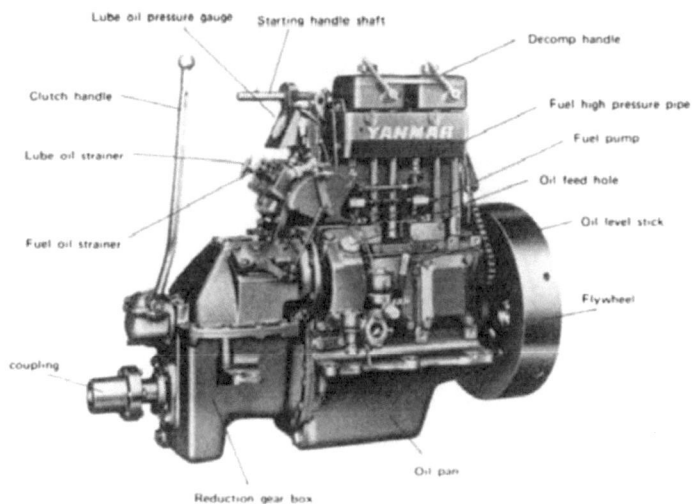

EXTERNAL DIMENSIONS (3SM)

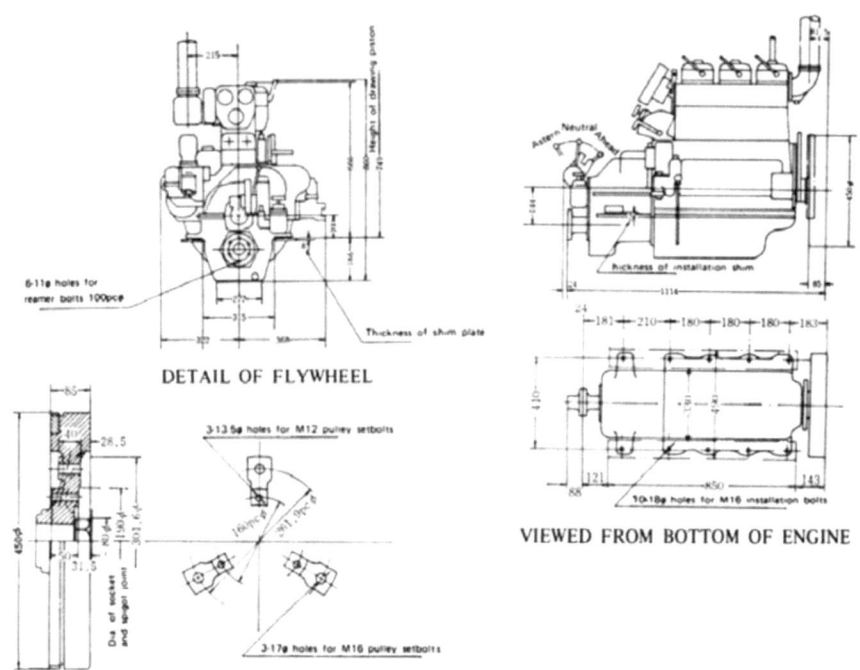

DETAIL OF FLYWHEEL

VIEWED FROM BOTTOM OF ENGINE

PARTS LOCATION (3SM)

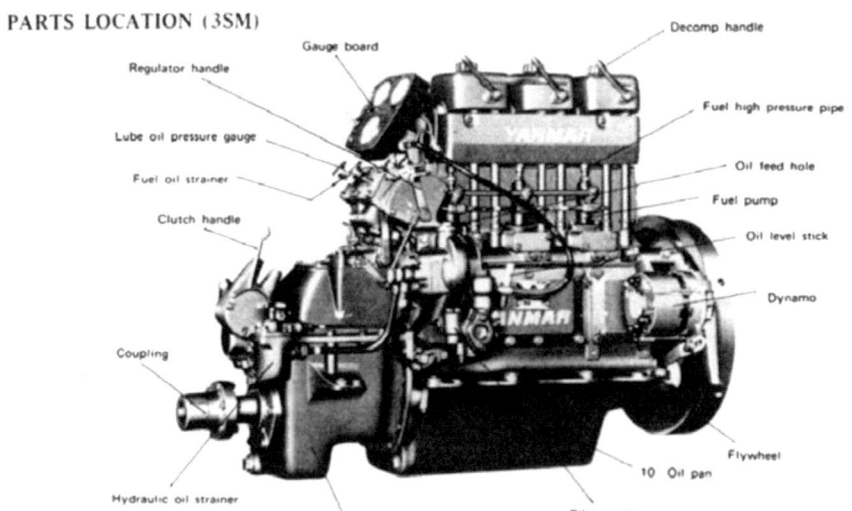

PREFACE

To operate the engine always under its best conditions, remember the following five points:

★ The engine always requires clean fuel oil of the best quality.
  Always keep the fuel tank, strainers, and fuel pipes clean.
★ The engine is always in need of clean lubricating oil of the best quality.
  Use an adequate grade lubricating oil and maintain the oil level above the minimum line on the oil level gauge at all times.
★ The engine must always have clean air.
  Check that there are no carbon deposits or other foreign particles precipitated at the air intake ports and exhaust system.
★ The engine must always be water cooled.
  Supply a sufficient amount of cooling water in the cooling system.
★ The engine functions more efficiently under normal load conditions.

1. OUTLINE OF STRUCTURES

No.	Division	Parts	Description		
			1SM	2SM	3SM
1	Engine body	Cyl. block	Monoblock piece for water jacket and crank case		
		Cyl. liner	Separate piece from cylinder, wet-type		
		Main bearing	Metal housing	Lift up from cylinder block	
		Oil sump	Cyl. block unit	Oil pan	
2	Suc. & exh. device & valve meachanism	Cyl. head	Water cooling of cylinder individually		
		Suc. & exh. valves	Mushroom-shape, seat angle of 45°		
		Exh. manifold	Bend-type air cooling insulating material	Exh. manifold type air cooling, w/o insulating material	Exh. manifold type water cooling
		Exh. silencer	Circle shape no-resistance type		
		Air cleaner			
		Valve mechanism	Valve push rod, suspending arm type, suc. & exh. cam w/baffer curve line		
3	Main working section	Crankshaft	Stamp forging		
		Flywheel	Fitted from crankshaft and taper part, w/power take-off pulley		
		Piston	Oval process, anti-abrasive ring set top ring cut.		
		Piston pin	Floating type		
		Piston rings	3 pressure rings and 2 oil scraper rings		
		Connecting rod	I section, stamp forging		
		Crankpin metal	Thick metal	Thin metal	
4	Lub. system	Lub. oil pump	Spur gear type		
		Lub. oil strainer	Full flow type, auto-clean type strainer	Full flow type, auto-clean type strainer	Full flow type, by-pass centrifugal strainer
		Oil level gauge	Plug-in type		
		Lub. oil cooler	Water-cooling single-pipe type	Water-cooling single-pipe type	Water-cooling multi-pipe type
5	Cooling system	Heat exchanger			
		Plunger pump	Single-acting plunger type		
6	Bilge system	Plunger pump	Single-acting plunger type		
7	Fuel system	Fuel injection pump	Bush type, independent cylinder		
		Fuel injectio valve	Pintle valve		
		Fuel strainer	Auto-clean type		
		Fuel tank	Made of steel board		
8	Governor system	Governor	Centrifugal, all-speed type		
9	Starting system	Manual starting			
		Chain starting	Accelerating starting by chain		
		Electric starting		(special order)	(standard equipped)
		Air starting			
10	Reducer & clutch system	Reducer	Spur gear type		
		Reversing clutch	Wet-type, single plate, mechanical	(hydraulic-special order for 2SM)	Wet-type, single plate, hydraulic oil type
11	Remote control system	Governor remote control system	Wire-type remote control		
		Clutch remote control system	Wire-type remote control		
12	Engine bed	Wooden bed, etc.			

2. FUEL AND LUBRICATING OIL

As you already know that there are many classes and grades of fuel and lubricating oil marketed today for use in diesel engines. Naturally, a selection of wrong class and/or grade of fuel or lubricating oil might result in unexpected trouble of a diesel engine or otherwise sure to shorten the serviceable life of the engine. Use of quality fuel and lubricating oil of right class and grade will increase the life of the engine many times, offsetting the higher price of the quality oil because on long terms long serivce of the engine gives its owner much more savings than use of low-cost oil which tends to shorten the engine life.

2-1. Fuel
Except gas engine, nearly all types of internal combustion engine burn fuel derived from petroleum for power source. In the following paragraphes, light oil and heavy oil used as fuel to run diesel engines will be explained.

2-1-1. Light Oil
1. Diesel Light Oil
   In general, light oil has the specific gravity of 0.83~0.89 and the boiling point of 200°C~370°C. Diesel light oil is widely used to run high-speed diesel engines of 1200 rpm or more employed in agricultural machinery, automobile, construction equipment, etc.
2. Requirements of Diesel Light Oil
   1) High cetane rating:
      Good ignitability and high combustion efficiency Generally these requirements are met diesel light oil of the cetane rating of over 45.
   2) Low sulphur content:
      High sulphur content of the oil speeds up corrosion and wear of engine parts particularly those parts which directly come in contact with fuel. For these reasons, such oil should not contain 1% or more of sulphur.
   3) Appropriate viscosity:
      Degree of viscosity must be appropriate with relation to ignition and combustion. If the viscosity is too high, atomized fuel particles are too large for dispersion; thus, the combustion time lags and color of the exhaust gas becomes poor. On the other hand, if the viscosity is too low, the atomized particles are too small for penetration of injection, resulted in seizure of the plunger and injection nozzle as they are not provided with lubricating action.
   4) No mixed dust and moisture:
      Impure oil usually contains dust and moisture which cause damage to the plunger and injection nozzle. Use of pure fuel is highly recommended. Besides, be sure to filter the fuel prior to supply to the engine.

3. Cetane Rating

Cetane rating is the most conveniently used criterion for rating diesel fuel and is equivalent of octane rating for gasoline. Cetane rating is used as index of the ignitability and refractoriness. Low cetane rating fuel has poor ignitability and tends to cause diesel knock. Causes of diesel knocking are in many cases opposite to those of knocking of gasoline engine, as can be seen in the following table.

Causes of Engine Knocking

		Gasoline Engine	Diesel Engine
Engine	Compression Ratio	High	Low
	Temperature & Pressure of Suction Air	High	Low
	Temperature of Cylinder Wall	High	Low
	RPM	Low	High
	Ignition Point of Fuel	Low	High
	Cylinder Capacity	Large	Small
Fuel	Octane Number	Low	———
	Cetane Number	———	Low

2-1-2. Properties of Fuel & Engine Performances

Property of Fuel	Starting Characteristic	Smoothness of Operation	Smoke Generation	Exhaust Fume	Output	Fuel Consumption	Accumulation within Combustion Chamber
Ignitability (Cetane Rating)	No direct relation but higher cetane rating, better starting characteristic.	No direct relation but higher rating, better smoothness of operation.	Closely related; not much difference if cetane rating is above the min. cetane rating limit.	Direct relation; higher cetane rating, lower exhaust fume.	No relation.	No relation.	Relation exists; higher cetane rating, lower the accumulation.
Volatility (90% end point)	No clear relation.	Relation exists; poorer volatility, better the performance.	Direct relation; better volatility, higher smoke generation.	No clear relation.	No relation.	No relation.	Relation exists; smaller this property, larger this aspect.
Viscosity	No clear relation.	Relation in a certain degree, high viscosity is not good.	Relation exists; higher viscosity, dependant on relation with the higher volatility.	Not related specifically.	Not related.	Not related.	Direct relation exists; dependant on relation with volatility.
Specific Weight	Not related.	Not related.	Direct relation exists; dependant on relation with volatility.	Not related specifically.	Direct relation exists; dependant on relation with calorific value.	Direct relation exists; dependant on relation with calorific value.	Relation exists; dependant on engine characteristic.
90% Residual Oil & Carbon Content	Not related.	Not related.	Relation exists; slightly inverse relation.	Not related specifically.	Not related.	Not related.	Relation exists; slightly inverse relation.
ASTM Gum	Not related.	Not related.	Relation exists; slightly inverse relation.	Not related specifically.	Not related.	Not related.	Relation exists; slightly inverse relation.

Property of Fuel	Starting Characteristic	Smoothness of Operation	Smoke Generation	Exhaust Fume	Output	Fuel Consumption	Accumulation within Combustion Chamber
Sulphur Content				Not related specifically.			
Flash Point				Not related specifically.			

2-1-3. Cautions on Fuel

It is important to select the fuel which does not contain dust and water vapor particularly. Besides, prior to filling the fuel tank, filtering is required. To filter fuel, use a piece of clean, finely interwoven cotton or the like cloth. Fuel come in drum can contains some amount of impurities such as dust and water particles settled down on its bottom. Refrain from inverting the can prior to transferring fuel therefrom or avoid pumping up fuel from the very bottom of the can.

2-2. Lubricating Oil

Lubricating oils presently used for all types of internal engines are about all of them mineral oil refined from petroleum. Depending upon their application, viscosity and quality (superior, regular, with/without additive), there are numerous kinds marketed today. Among them are lubricants for diesel engine use. We shall describe the appropriate kinds, property requirements and handling method of this group of lubricants shortly.

2-2-1. Engine Oil

1. Purpose of Engine Oil Usage

   Engine oil mainly purports to check on friction and wear-off of clearance between the cylinder wall and piston rings and bearing section of the pin and journal portions of crankshaft. Besides, not only it seals off a gap between the cylinder wall and piston rings and thereby prevents blow-by of combustion gas and consequently rules out decrease of produced output, but also remove harmful impurity from various sections of the engine, playing a role of preventing corrosion and rusting as well as of carrying away and cooling the heat due to friction.

2. Types of Engine Oil

   Engine oil is roughly classified into two groups; namely, motor oil (gasoline engine oil) and diesel engine oil. Within respective group, they are further classified into several types, depending upon quality, usage condition, and viscosity.

Classification by Usage Condition
(API Service Classification)

Classification	Symbol	Quality & Application
Diesel Engine Use	DG	Used under the light-load operational condition of diesel engine. Requirement is small in checking on producing of worn-off particles and abrasion due to designed roles of fuel, lubricating oil and engine itself.
	DM	Used under the severe operational condition of diesel engine when producing of worn-off particles, abrasion, etc. are evident due to sulphur content of fuel used or when residual carbon of lubricating oil affects greatly upon engine design.
	DS	Used under the heavy-load operational condition of diesel engine when the fuel producing worn-off particles to a great extent, or when abrasion of the engine especially designed for high-output, high-load operation, etc. is serious.

NOTE: *Symbols operated in above table have the following meanings.*

	Symbol	Meaning
First Letter	M	Motor
Second Letter	M	Moderate
	S	Severe
	G	General

Engine oil has been refined and blended with addition of such necessary additives as oxidation inhibiter, corrosion inhibiter, rust preventive, dispersant, etc.

3. SERVICING AND PERIODICAL CHECKING

3-1. Main points of servicing and training

Daily servicing and a periodical checking are necessary to keep the engine always in good conditions. At the time of delivering the engine, it is advisable to explain to each user about handling and servicing procedures satisfactorily by using the service manual in which are given the main points of handling and periodical checkings.

	Treatment	Instruction to engine operator	Reason
	Selection and storage of fuel oil	Use approved diesel oil listed on Page 99. Avoid exposing the oil to the sun. Keep away from dust. Strictly avoid using sedimentary oil.	
	Selection and storage of lub. oil	Use approved lubricating oil listed on Page 99. Avoid exposing it to the sun. Keep away from dust.	Inferior quality oils are apt to cause quick abrasion and damage.
Preparation prior to use	Check each nut and bolt.	Check leakage of oil or water, and clogging.	
	Retighten, if necessary.	Retighten, if necessary.	
	Check and supply engine oil.	Supply oil until it reaches the F-mark of oil level stick.	Being out of oil causes seizure in moving parts.
	Check and supply fuel oil.	Avoid dust. Do not use sedimentary oil.	Foreign matters cause damage and wear of parts of fuel system.
	Supply oil to each part.	1. Fuel injection pump regulating rack. 2. Plunger parts of cooling water pump and bilge pump. 3. Pinion shaft of cell motor. 4. Moving part of remote control device; shaft part of handle.	To help smooth slide on each part and to prevent from seizure.
	Handle operation of fuel and lub. oil strainers.	Turn the handles a few times.	To clean clogged elements.

— 8 —

	Treatment	Instruction to engine operator	Reason
Operation	Hand operation before strainers	Set the regulator handle at the "Stop" position, allowing no compression, and turn the flywheel by hand about ten times.	Permeate oil through each part to attain smooth operation. Confirm if no abnormal sound being made in each part.
	Turning	Set the reversing handle at the "Neutral" and the regulator handle at the "Stop" positions, allowing no compression, and then push the starting, switch to turn the cell motor about ten rounds.	
	Priming (confirmation of fuel injection)	Set the regulator handle at the position "Run" and move it from side by priming handle four or five times. Then injection sound "betzu, betzu, . . . " can be heard. When no such sound heard, ventilate the fuel injection device.	
	Use of gasoline in cold season	When starting is extremely difficult, supply gasoline to the suc. inlet cover. Do not supply too much.	
Caution in operation	Confirmation of oil pressure	Ahead, by 1500 rpm; 1. Lub. oil pressure is $2 \sim 2.5$ kg/cm^2. 2. Hydraulic oil pressure is $9 \sim 9.5$ kg/cm^2.	Refer to page 77.
	Warming-up	Practise no-load operation by $700 \sim 800$ rpm for more than five minutes to warm up the enigne.	To attain thorough oil permeation. To prevent the liner piston from seizure due to sudden increase in temperature immediately after starting.
	Confirmation of cooling water	Confirm whether the circulation of cooling water is normal or not.	Prevention of seizure.
	Confirmation of lub. oil	Remove the bonnet, and confirm whether the circulation of lub. oil is normal or not around valve lever shaft and valve arm guide.	Prevention of seizure.

	Treatment	Instruction to engine operator	Reason
Caution in operation	Observation in color exhaust gas	1. Colorless to blueish white; the engine in the best condition. 2. Black; a trouble in the engine or overload. 3. White; lub. oil is burning. 4. Continuous exhaustion of black smoke should be avoided.	Refer to page 94.
	Check abnormal sound or abnormal increase in temperature	When abnormal sound or abnormal increase in tmeperature is observed, stop the operation immediately and check the causes.	If not checked at the time, the damage may become bigger.
	Check gas or water leakage.	Check gas or water leakage, and retighten bolts and nuts, if necessary.	
	Avoid resonance.	By certain number of revolution, resonance between the engine and the hull may occur depending upon the structure of hull.	Avoid using the engine by about the number of revolution which may cause resonance; otherwise undesirable effect on the engine and the hull occurs.
	Check charge lamp.	If charge lamp does not go off even after high speed operation reached, charging circuit is in trouble.	
	Supply fuel oil.	Supply oil while the oil level can be seen in the oil level gauge of fuel tank.	In case that the engine stops due to running-out of oil, ventilation of fuel pump is necessary after oil supplied.
Engine stop	Practise idle operation before the engine stops.	Practise idle operation for about five minutes by setting the clutch at the "Neutral" position. Give high speed operation momentarily before stop the engine.	Discharge carbon in the combustion chamber out of the engine.
	Stop the engine by using the regulator handle.	Use the regulator handle to stop the engine.	If decomp. handle is used to stop the engine, imperfectly combusted fuel oil will be accumulated in the piston, causing carbon clogging in the clearance between valve and valve seat which incurs difficulty in next starting.

	Treatment	Instruction to engine operator	Reason
Service after used	Drain cooling water.	After using the engine, drain cooling water by opening the cocks of cooling water pump and exhaust manifold. Complete drainage is obtained by turning the flywheel. In cold season, tighten the Kingstone cock.	In cold weather, frozen water in the cylinder may damage the cylinder.
	Clean the engine.	While the engine is still warm, clean dust and dirt completely.	For better maintenance of engine.
Storing engine	Change lub. oil.	Change lub. oil and allow to turn a few times.	For next use. Treatment for rust-proof.
	Keep the outer parts away from corrosion and dust.	Wipe dust and dirt away. Apply oil to the following parts: 1. Link part of fuel pump. 2. Pinion shaft of cell motor. 3. Moving part of remove control device, shaft part of handles. Cover the engine with vinyle, paper, or cloth for dust-proof.	
	Occasionally turn the engine.	Turn the engine by hand once a month.	For rust-proof of metal, piston, liner rings.
	Store battery.	Disconnect the wire. Store in dry place. Charge once a month.	

3-2. Time of Periodical Checking

A periodical checking is necessary to keep the engine in good conditions at all times. The frequency of periodical checking may vary depending upon the purpose of using the engine, conditions of use, quality of oils to be used, and methods of handling the engine, and it is difficult to generalize on the frequency of periodical checking and servicing. Therefore, general explanations will be given here. The relationship between the detail of checks and time is as follows:

	Checking & Servicing Item	Daily	Every 50 hr.	Every 250 hr.	Every 500 hr.	Every 1000 hr.	
Fuel Oil Cleaning of fuel strainer	Checking the level & supplying of fuel oil.	○					
	Discharge of drain from fuel strainer	○	(Prior to supply)				
	Discharge of drain from fuel strainer	○	○				
	Turn handle of fuel strainer	○					
	Cleaning of fuel strainer				○		
Lub. Oil	Checking crank case & oil level of clutch	○					
	Removal of drain from lub. oil strainer (engine & clutch sides)		○				
	Handle of lub. oil strainer (engine & clutch sides)		○				
	Cleaning of lub. oil strainer (engine & clutch sides)						
	Cleaning of centrifugal strainer			○			
	Cleaning of lub. oil cooler						3000 hr.
	Change of lub. oil			○			
Cooling water pump Gland packing	Checking water leakage (Retightening in case of water leakage)	○					
	Disassembling & Checking					○	
Bilge pump Gland packing	Checking water leakage (Retightening in case of water leakage)	○					
	Disassembling & Checking					○	
Fuel injection pump	Check of injection timing				○		
	Disassembling & checking of main part						2000 hr.
Fuel injection valve	Check of injection				○		
	Cleaning of strainer				○		
Cylinder head	Retightening of clamp bolts	100 hrs after trial operation of restration					
	Cleaning of combustion chamber				○		
	Cleaning of pre-combustion chamber				○		
	Adjustment of valve clearance (suc. & exh. valves)				○		
	Grinding of suc. & exh. valves				○		
Piston	Disassemblying. Check of ring					○	
Checking anti-corrosive zinc					○		
Oiling	Oiling to cooling water pump	○					
	Oiling to cell motor pinion shaft	○					
	Oiling to speed regulator assembly	○					
Clutch	Disassembling. Checking. Cleaning						5000 hr.
	Exchange of rubber O-ring						5000 hr.
Adjustment of remote-control wire		500 hrs after fitting					
Check of dynamo drive belt tension				○			

4. ENGINE DISASSEMBLY

4-1. Preparation for disassembly

Prior to disassembling engine, the following precautions should be observed.
1. A clean, dust-free workshop should be availabe for the disassembly.
2. Prepare a table or board and a tin can by which the disassembled parts can be prevented from damage or losing before reassembly.
3. Prepare a washer and a washing tin can.
   In cleaning the disassembled parts in a shop, prepare a washer or a washing tin can being sold on the market. In cleaning the parts on a ship, prepare a 18ℓ-tin cut into two parts as in the figure below.

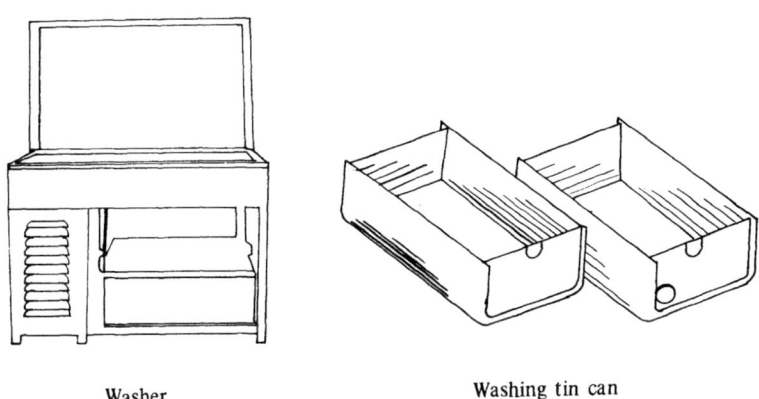

Washer　　　　　　　　Washing tin can

(4) Disassembly Tools

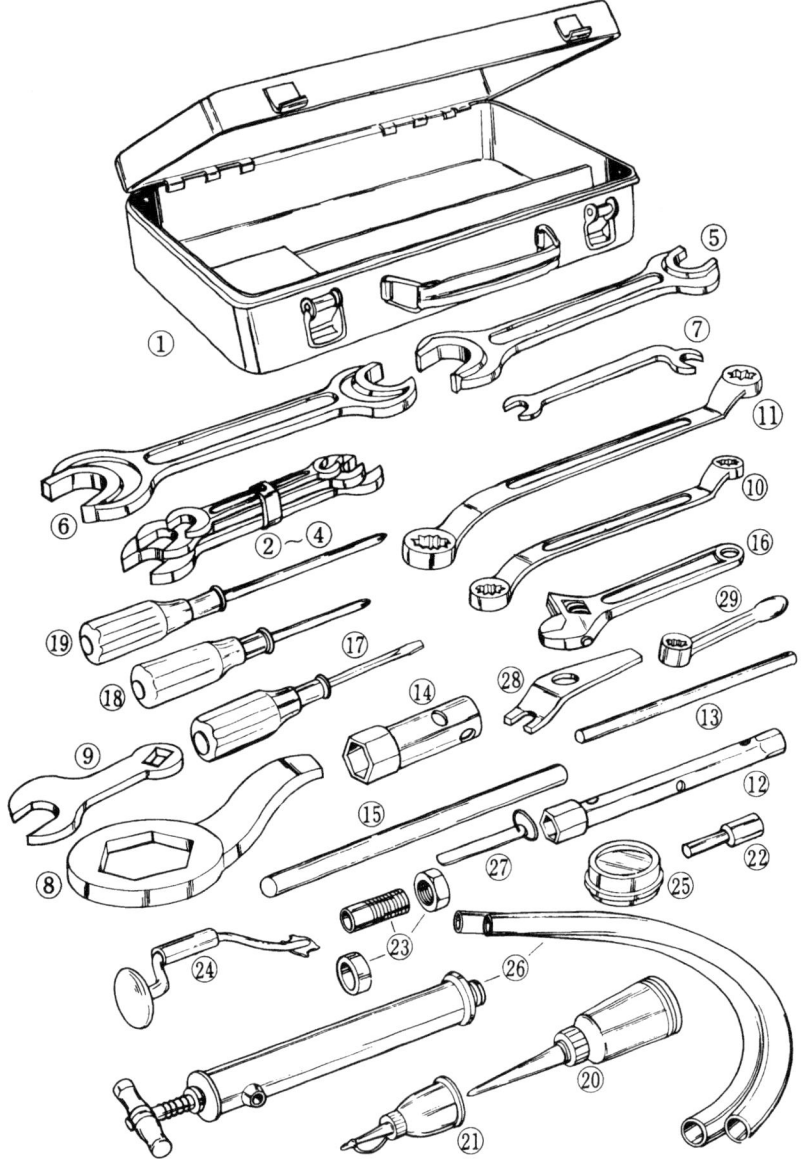

Standard Disassembly Tools

Figure Number	Name of Tool	Figure Number	Name of Tool
1	Tool box	15	Handle 16ϕ x 300
2	Double-ended spanner 10 x 14	16	Monkey wrench 200
3	Double-ended spanner 17 x 19	17	Screwdriver 6 x 100
4	Double-ended spanner 21 x 23	18	Plus screwdriver 6 x 100
5	Double-ended spanner 26 x 29	19	Plus screwdriver 8 x 150
6	Double-ended spanner 32 x 35	20	Oil feeder
7	Special spanner 10 x 14 (For governor link)	21	Gasoline feeder
		22	Strainer nozzle removing tool
8	End-nut spanner	23	Delivery valve guide removing tool
9	Special spanner (17) (For fuel injection pump adjusting lock nut)	24	Valve lapping tool
		25	Lapping powder
10	Double offset wrench 14 x 17	26	Hand pump (with vinyl pipe)
11	Double offset wrench 21 x 23	27	Thickness gauge 0.15
12	Box spanner 14 x 17	28	Suc./exh. valve handling tool
13	Handle 8ϕ x 200	29	Priming handle
14	Box spanner (29) (For head bolt) (For main bearing tightening nut)		

Special-order Disassembly Tools

Name of Tool	Code Number	Name of Tool	Code Number
Liner removing tool ass'y	722310–92130	Plier (for hole) (for circlip (10–40ϕ))	28190–000120
Removing tool (for cam gear, flywheel)	122510–92400	Plier (for hole) (for circlip (40–70ϕ))	28190–000130
Bar spanner for hex. socket bolt (for ring gear)	102302–92750	Removing tool (A) (for reduction pinion)	122510–92800
Piston inserting guide	122310–92140	Removing tool (B) (for reduction pinion)	122510–92810
Special box spanner (for cam shaft nut)	122510–92710	Hex. spanner (for oil pressure working cylinder tightening bolt)	122510–92820
Plier (for shaft) (for circlip (11–35ϕ))	28190–000020		

4-2. Precautions for Disassembly

The following cautions should be observed for more efficient and assured disassembling.
1) Disassemble the engine according to the orders of disassembling.
2) Do not disassemble any parts unnecessarily.
3) Clean the exterior surface of the engine.
4) Use right tools only; otherwise damage of parts may occur.
5) Pay attention not to drop the disassembled parts or hit each others.
6) Place the disassembled parts in a right order.
7) Remember where and how they were mounted.
8) Place the removed cotter pins, nuts, bolts and washers beside their parts or at the place where they were.
9) Confirm the match-marks of each part.
10) Check if there are any damaged or worn parts.

4-3. Disassembling Procedure

	Part Name	Piece	Order and Caution	Tool Used	Figure
1.	Exh. silencer	1	Remove the exh. pipe and exh. silencer.		
2.	Suc. inlet cover	1	Remove it.	Box spanner (14) Plus screwdriver	
3.	Fuel injection valve High pressure fuel pipe	3	Remove them.	Spanner (19)	
	Fuel overflow pipe (fuel pump side)	1	Remove it.	Box spanner (17)	

Part Name	Piece	Order and Caution	Tool Used	Figure
Fuel overflow pipe (fuel injection valve side)	1	Remove it.	Spanner (17)	
Fuel injection valve	3	Remove the fuel valve holder and then fuel injection valve.	Box spanner (17)	
4. Exh. manifold Cooling water by-pass pipe (for bilge pump)	1	Loosen it.	Box spanner (17)	
Cooling water pipe (Cyl. head – exh. pipe)	3	Remove them.	Box spanner (14)	
Exh. manifold	1	Remove it.	Spanner (17) Double offset wrench (17)	
5. Cylinder Head and Bonnets	3	Remove them.	Box spanner (17)	

Part Name	Piece	Order and Caution	Tool Used	Figure
Suc./exh. valve lever support ass'y	3	Remove the sux./exh. valve lever support ass'y.	Double offset wrench (23)	
Bonnet ass'y		Remove them.	Box spanner (17)	
Lub. oil pipe (cyl. head − cyl.)	1	Remove it.	Spanner (17)	
		Loosen lub. oil overflow pipe.	Spanner (19)	
Cyl. head ass'y	3	Remove the cyl. head. Remove the packing (for cyl. head).	Box spanner (29) Torque wrench (29)	
Valve push rod	6	Remove the rods and the rod covers.		
Valve push rod cover	6	Refer to page 61.		
Cyl. liner	3			

− 18 −

	Part Name	Piece	Order and Caution	Tool Used	Figure
6.	Gauge Board Hydraulic oil pipe (hydraulic oil strainer ~ pressure gauge)	1	Remove it.	Spanner (17)	
	Tachometer flexible shaft	1	Remove it.	Double offset wrench (23)	
	Lub. oil pipe (cyl. ~ pressure gauge)	1	Remove it.	Spanner (17)	
	Gauge board ass'y	1	Remove the gauge board ass'y (including engine lifting metal)	Spanner (17) Spanner (19)	
7.	Fuel Injection pump Fuel oil pipe (strainer ~ pump)	1	Remove it.	Spanner (21)	
	Governor lever shaft connectiong pin	1	Remove it.	Minus screwdriver Pliers	
	Fuel injection pump	3	Remove them.	Spanner (14)	

	Part Name	Piece	Order and Caution	Tool Used	Figure
	Priming shaft holder	3	Remove them.	Plus screwdriver	
	Priming shaft	3	Remove them.	Pliers	
	Roller guide	3	Remove them.		
8.	Regulator Handle Support				
	Lock bolt for governor link	1	Remove it.	Spanner (10)	
	Regulator handle support	1	Remove it.	Box spanner (14)	
9.	Governor Housing Ass'y	1	Remove it.	Spanner (14)	
10.	Fuel & Lub. Oil				
	Lub. oil pipe (cyl. outlet ~ centrifugal strainer)	1	Remove it.	Spanner (14) Box spanner (14)	
	Lub. oil pipe (cyl. outlet ~ strainer inlet)	1	Remove it.	Spanner (14) Box spanner (14)	
	Lub. oil pipe (strainer outlet ~ cooler inlet)	1	Remove it.	Spanner (14) Box spanner (14)	
	Union	1	Remove it.	Spanner (26)	

	Part Name	Piece	Order and Caution	Tool Used	Figure
	Fuel, lub. oil strainer ass'y	1	Remove it.	Spanner (17) Box spanner (17)	
11.	Charging Dynamo	1	Loosen the V-belt adjuster.	Spanner (17) Spanner (14)	
		1	Remove the charging dynamo.		
12.	Centrifugal Strainer Centrifugal strainer	1	Remove it.	Spanner (14)	
	Centrifugal strainer support	1	Remove it.	Spanner (19)	
13.	Starting Motor	1	Remove it.	Spanner (19)	

Part Name	Piece	Order and Caution	Tool Used	Figure
14. Cylinder Side Cover Breather	1	Remove it.	Spanner (17)	
Cyl. side cover (A) Cyl. side cover (B) Cyl. side cover (C)	1 1 1	Remove all of them.	Box spanner (17)	
15. Piston & Connecting Rod Connecting rod bolt	6	Unbend the bend washer by chisel and remove the connecting rod bolt.	Chisel Hammer Spanner (21) Double offset wrench (21)	
Connecting rod large-end cap	3	Remove the connecting rod large-end cap together with crank pin metal. Set the piston at the top dead center.		
Piston & connecting rod ass'y	3	Remove them. (Note) *Lightly push the crank pin metal part forward with the handle of hammer.*		
16. Cooling Water Pump Cooling water pump (water pump outlet ~ cooler inlet)	1	Remove the cooler side bolt of cooling water pipe (water pump ~ cooler inlet).	Double offset wrench (14)	

	Part Name	Piece	Order and Caution	Tool Used	Figure
	Cooling water pump	1	Remove it. (Note) *Pay attention not to damage the plunger.*	Double offset wrench (17) Wooden hammer	
17.	Lub. Oil Cooler Lub. oil pipe (cooler outlet ~ cyl. inlet)	1	Remove it.	Double offset wrench (14)	
	Lub. oil cooler ass'y	1	Remove it.	Box spanner (17)	
	Lub. oil overflow pipe	1	Remove it.	Box spanner (19)	
	Lub. oil pipe (cyl. ~ cyl. head)	1	Remove it.	Double offset wrench (21)	
18.	Bilge Pump Ass'y	1	Remove it. (Note) *Pay attention not to damage the plunger.*	Spanner (17)	

Part Name	Piece	Order and Caution	Tool Used	Figure
19. Hydraulic Oil Strainer Hydraulic oil pipe (hydraulic oil strainer ~ switch valve)	1	Remove it.	Double offset wrench (21) Double offset wrench (23)	
Hydraulic oil pipe (hydraulic oil pump ~ hydraulic oil strainer)	1	Remove it.	Spanner (26) Spanner (32)	
Hydraulic oil strainer ass'y	1	Remove it.	Double offset wrench (17)	
20. Rear Cover Switch Valve Hydraulic oil pipe (rear cover ~ intermediate shaft)	1	Remove it.	Spanner (17)	
Wire socket (for remote control)	1	Loosen it. Loosen the clutch upper case tightening bolt.	Plus screwdriver Box spanner (17)	
Rear cover switch valve ass'y	1	Remove it.	Box spanner (17)	
21. Cylinder Gear Box Side Cover Hydraulic oil pipe (cyl. outlet ~ hydraulic oil pump)	1	Remove it.	Double offset wrench (17) Spanner (26)	
Hydraulic oil pump ass'y	1	Remove it.	Spanner (14)	
Hydraulic oil pump driving gear	1	Remove it.	Box spanner (14)	
Cyl. gear box side cover	1	Remove it.	Spanner (17)	

	Part Name	Piece	Order and Caution	Tool Used	Figure
22.	Reduction Reversing Gear Upper Case Lub. oil pressure	1	Remove it.	Spanner (17)	
	Reduction reversing gear upper case	1	Remove it.	Spanner (19) Spanner (17)	
23.	Governor Weight Slide barrel Governor weight	1 1	Remove it. Remove it.	Minus screwdriver Pliers	
24.	Clutch Housing	1	Cut the wire and take out the clutch housing.	Pliers Double offset wrench (17)	
25.	Cooling Water Pump Plunger & Bilge Pump Plunger Cooling water pump plunger	1	Remove the circlip. (Note) *Do not lose the circlip.*	Radio pliers	
			Pull out the plunger pin by using 8mm-bolt.	8mm-bolt	

Part Name	Piece	Order and Caution	Tool Used	Figure
		Remove the cooling water pump plunger.		
Bilge pump plunger	1	Remove the circlip in the same manner to cooling water pump plunger. Pull out the pin by using the 5mm-bolt. Remove the plunger.	Radio pliers 5mm-bolt	
Cam shaft nut	1	Unbend the washer, loosen shake proof washer and remove the cam shaft nut.	Special box spanner	

26.

Part Name	Piece	Order and Caution	Tool Used	Figure
Intermediate Gear Shaft Tightening nut	1	Unbend the shake proof washer and unscrew the tightening nut.	Minus screwdriver Hammer	
Washer for intermediate gear	1	Unbend the bend washer and remove the washer for intermediate gear.	Minus screwdriver Hammer Spanner (14)	
Intermediate gear Intermediate gear shaft	1 1	Hammer out the intermediate gear shaft from the outside to the inside by using wooden hammer.	Wooden hammer	

	Part Name	Piece	Order and Caution	Tool Used	Figure
27.	Reduction Reversing Gear Lower Case	1	Remove it.	Spanner (17)	
28.	Thrust Shaft Tightening nut	1	Unbend the shake proof washer and remove the tightening nut.	Special box spanner	
	Shake proof washer	1	Remove the shake proof washer and the thrust shaft bearing washer.		
	Thrust shaft bearing washer	1			
	Thrust shaft rear cover	1	Remove the hex. bolts and the thrust shaft rear cover.		
	Hex. bolt	6			
	Thrust shaft	1	Hammer out the thrust shaft from the inside to the outside of the reversing gear case by using wooden hammer.	Wooden hammer	
29.	Flywheel End-nut flywheel	1	Unscrew the end-nut and remove the lock washer.	End-nut spanner Hammer	
	Flywheel		Detach the flywheel by using flywheel removing tool. (Note) *Set the flywheel removing tool in parallel with the end surface of crank shaft.*	Flywheel removing tool Double offset wrench	

— 27 —

	Part Name	Piece	Order and Caution	Tool Used	Figure
	Flywheel		(Note) *After the removing bolt is fully screwed in the screw hole of flywheel, then tighten the nut.*		
30.	Camshaft				
	Camshaft nut (flywheel side)	1	While fixing the camshaft by applying a rod to the camshaft gear, remove the camshaft nut (flywheel side).	Spanner (32)	
	Dynamo pulley	1	Remove it.		
	Camshaft cover	1	Remove it.	Spanner (10)	
	Camshaft gear	1	Push camshaft toward the flywheel side by using camshaft gear removing tool, and then remove the camshaft gear, cooling water pump connecting rod and tachometer driving gear.	Camshaft gear removing tool	
	Cooling water pump connecting rod	1		Double offset wrench	
	Tachometer driving gear	1			
	Tappet	6	While pushing out the camshaft, pull out the tappets.		
	Camshaft	1	Remove the camshaft. (Note) *In removing the camshaft, pay attention not to damage the camshaft.*		
	Starting motor bracket	1	Remove it.	Spanner (21)	

	Part Name	Piece	Order and Caution	Tool Used	Figure
31.	Crank Shaft		Drain lub. oil from the lub. oil tank and run the engine reversely.		
	Oil pan	1	Remove it.	Double offset wrench (17)	
	Cylinder front cover	1	Remove it.	Double offset wrench (17)	
	Lub. oil pipe (cyl. outlet ~ strainer inlet)	1	Remove it.	Spanner (14)	
	Oil pan	1	Remove it.		
	Lub. oil pump	1	Remove it.	Double offset wrench (17)	
	Slotted nut for main bearing holder	8	Remove the wire and the slotted nut.	Pliers Torque wrench (26)	
	Main bearing holder (gear side)	1	Remove the main bearing holder.		
	Main bearing holder (intermediate part)	2			
	Main bearing holder (flywheel side)	1			
	Main bearing metal (gear side)		Remove the main bearing		
	Main bearing metal (intermediate part)				
	Main bearing metal (flywheel side)				
	Thrust bearing metal	2	Remove the metal. (Note) *Remember the direction of metal.*		
	Crank shaft	1	Remove the crank shaft.		

5. ENGINE REASSEMBLY

5-1. Precautions for Reassembly

1. When cleaning the disassembled parts, pay attention to the following points.
   1) Wash the precision parts such as fuel injection pump and fuel injection valve with new washing oil.
   2) Washing is carried out in such an order as from more precisely fabricated parts to less.
   3) Particularly sliding parts should be cleaned well.
   4) After cleaning, do not wipe the precision parts and sliding parts with rag.
2. When reassembling the engine, pay attention to the following points.
   1) Wash the parts once again before reassembling.
   2) Apply oil to sliding parts.
   3) Use brand new bend washers, split pins, and paper packings.
   4) Apply proper strength to tighten bolts, screws and nuts.
   5) When parts are fitted with several bolts, screws or nuts, tighten them evenly in diagonal order. Should these parts are tighten unevenly, they are becoming the cause of damaging bolts, screws or nuts, and of leakage of oil, water and gas. Be sure to use washers where the washers (flat washer, spring washer and lock washer) were used.
   6) Be sure to match the setting marks of such parts as crankshaft gear and connecting rod metal.
   7) Before reassembling parts, be sure to check if parts are in good condition.

5-2. Reassembling Procedure

	Part Name	Piece	Order and Caution	Tool Used	Figure
1.	Camshaft Tapet	6	Insert them applying mobil oil. Confirm whether working is lightly carried out or not.		
	Ball bearing #6207	1	Hammer ball bearing #6207 (gear side)in.	Wooden hammer	
	Bearing holding Plate	2		Spanner (10)	
	Hex. bolt M6 x 12	4	Fit the holding plates in correct direction.		
	Lock washer	2	Tighten the hex. bolt.		

Part Name	Piece	Order and Caution	Tool Used	Figure
Camshaft ass'y	1	Apply oil to camshaft metal.	Camshaft inserting tool	
Water pump connecting rod	1	Insert the set of the tachometer driving gear, camshaft gear, water pump connecting rod with camshaft inserting tool.		
Camshaft gear	1			
Tachometer driving gear	1	(Note) *The boss part of tachometer driving gear should be on the flywheel side.* (Note) *Pay attention not to damage the metal when camshaft is inserted. Revolution should be lightly carried out.*		
2. Crank Shaft				
Main bearing metal (upper)	4	Set the main bearing metals (upper) and apply some oil.		
Crank shaft ass'y	1	Assemble the crank shaft ass'y and crank shaft gear, match the setting mark (0–0), and then insert the crank shaft, ass'y. Apply some grease and insert the thrust metals.		
Crank shaft gear	1			
Thrust metal (upper)	2			
Main bearing holder (flywheel side)	1	Assemble main bearing holder by fixing thrust metal (lower) and main bearing metal (upper)		
Main bearing holder (intermediate part)	2			
Main bearing holder (gear side)	1			
Thrust metal (upper)	1			
Main bearing metal (lower)	4			
Main bearing holder	8	After temporarily tighten the main bearing holder nuts evenly, tighten firmly and alternately not to allow uneven tightening.	Torque wrench Pliers	

	Part Name	Piece	Order and Caution	Tool Used	Figure
	Wire 2φ	8	Tighten torque must be 17kg-m and crank shaft side-clearance should be 0.09~0.13. Back-lash should be 0.08~0.16. The direction of wire should be the same to the direction of tightening.		
3.	Lub. Oil Pump Lub. oil pump ass'y Driving gear Semicircular key Claw washer Hex. nut M14 Packing for lub. oil cyl. inlet Hex. bolt M10 Spring washer 10	1 1 1 1 1 2 4 4	Fix the driving gear to the pump body. Attach the pump body. (Note) *Before tightening, turn the crank shaft to match the gear teeth.*	Box spanner (17) Spanner (21)	
4.	Oil Pan Cylinder front part cover Cylinder packing Oil seal Hex. bolt M10 X 25 Straight pin	1 1 1 6 2	Hammer the straight pins in. Fit oil seal to cylinder front part cover and tighten with bolts.	Hammer Box spanner (17)	
	Lub. oil pipe (strainer ~ pump) Packing Hex. bolt M8 X 25 Spring washer	1 1 2 2	Insert lub. oil pipe into strainer of oil pan and tighten with bolts. Apply 3-bond on the bottom surface of cylinder and stick the packing. (Note) *Apply 3-bond uniformly.*	Spanner (14)	
	Oil pan ass'y Hex. bolt M10 X 28 Hex. bolt M10 X 63 Spring washer 10	1 17 2 19	Fix the oil pan ass'y and tighten the bolts evenly.		

	Part Name	Piece	Order and Caution	Tool Used	Figure
5.	Camshaft Nut Camshaft cover Camshaft cover, packing Oil seal #SB 30458 Hex. bolt M8 X 16	1 1 1 3	Attach the camshaft cover. (Note) *Be sure to place the oil escape groove at the bottom side.*	Spanner (14)	
	Dynamo pulley Camshaft nut (flywheel side) Washer Belt	1 1	Attach the dynamo pulley and tighten with camshaft nut.	Spanner (32)	
	Camshaft nut (clutch side) Shake proof washer	1 1	Tighten the camshaft nut on clutch side. (Note) *Avoid using minus screwdriver. Instead, use camshaft nut tightening spanner.*	Hammer Screwdriver Camshaft nut Tightening spanner	
6.	Cylinder Liner		Refer to page 62		
7.	Cooling Water Pump Plunger Plunger drip rubber Connecting rod pin Lock ring for hole 16 Cooling water pump ass'y Hex. nut (plated)M10 Cooling water pipe (water pump outlet ~ cooler inlet) Cooling water pump packing Hex. bolt (plated) M10 X 25	1 1 1 1 1 3 1 1 2	Insert water drip rubber. Insert plunger, setting the oil hole of connecting rod upwards. Lock the connecting rod with pin. Attach the cooling water pump (water pump outlet ~ cooler inlet) firmly to the pump. Attach the cooling water pump. (Note) *Use plated nut.*		

Part Name	Piece	Order and Caution	Tool Used	Figure
8. Piston & Connecting Rod		Refer to page about the assembly of piston and connecting rod.		
Piston (with ring)	3	Apply mobil oil to the outer surface of piston.	Piston inserting tool	
Piston pin	3			
Lock ring for hole 38	6	Arrange piston rings so that their cut parts deviate by 90° from next one.		
Connecting rod	3			
Piston pin metal	3	Insert piston. (Note) *Use inserting tool.*		
Connecting rod bolt	6	(Note) *To enable to tighten the rod bolts in the operation side, the "YANMAR" mark of connecting rod should be in the side of flywheel. While tightening, do not allow the large end of connecting rod to touch the pin part of crank shaft.*		
Lock washer	6			
Crank pin metal	3			
		Apply mobil oil to the crank pin part and tighten the large end part.		
		Tighten torque must be 9 kg-m.		
		Make sure of bending the washer firmly.		
9. Cylinder Head			Torque wrench	
Cyl. head ass'y	3	Fix the O-ring to the cooling water connecting pipe.	Box spanner (29)	
O-ring (No.1) P12	3		Solder	
O-ring	6	(Note) *Use new O-ring.*	Micrometer	
Cyl. head packing	3	Attach cyl. head with caution not to catch cyl. head packing.	Spanner (17)	
Cyl. head tightening bolt	12			
Pre-combustion chamber (front chamber)	3	Tighten the nut to the crossed direction with tightening torque 17kg-m.		
Pre-combustion chamber (rear chamber)	3	Confirm the top clearance. Put solder into the chamber hole and measure the clearance by means of micrometer.		
Pre-combustion chamber Packing A	3			

	Part Name	Piece	Order and Caution	Tool Used	Figure
	Pre-combustion chamber Packing B	3	Standard top clearance 2.35~2.55		
	Fuel injection valve	3	Assemble pre-combustion chamber in the order of packing (A), pre-combustion chamber (front chamber), packing (B) and pre-combustion chamber (rear chamber).		
	Fuel valve holder	3			
	Hex. nut 10	6			
			(Note) *Do not use wrong packings.*		
			Packing (A) bore 29ϕ		
			Packing (B) bore 35.5ϕ		
			Insert the fuel injection valves and fix the fuel valve holders, then tighten with hex. nuts.		
10.	Cylinder Side Cover		Attach the cyl. side covers. Insert the oil level stick.	Box spanner (17)	
	Cyl. side cover (A)	1			
	Cyl. side cover (A) packing	1			
	Cyl. side cover (B)	1			
	Cyl. side cover (B) packing	1			
	Cyl. side cover (C)	1			
	Cyl. side cover (C) packing	1			
	Hex. bolt M10×25	21			
	Oil level stick	1			
11.	Roller Guide & Priming Shaft		Insert the roller guides applying mobil oil.	Plus screwdriver (10)	
	Roller guide	3			
	Priming shaft	3			
	O-ring	3			
	Shaft holder	3			
	Fuel overflow pipe (fuel pump side)	1			
			Insert the priming shafts.		

	Part Name	Piece	Order and Caution	Tool Used	Figure
	Roller Guide & Priming Shaft		Attach the shaft holder and fuel overflow pipe at the same time.		
12.	Cell Motor Bracket Cell motor bracket Hex. bolt 12 X 45 Spring washer 12	1 3 3	Attach the cell motor bracket.	Double offset wrench (19)	
13.	Fuel Pump Fuel pump Fuel regulating rack adjusting shaft Spring for adjusting shaft Fuel injection regulating shim Hex. lock nut M4 Hex. nut M8	3 3 3 3 6 6	Assemble fuel pump, fuel regulating rack adjusting shaft, spring for adjusting shaft, fuel injection regulating shim, hex. lock nut. Attach the reassembled fuel pump to the cylinder and tighten with the nuts. After attaching, confirm whether the rack is lightly sliding by moving the rack by hand.	Spanner (14) Spanner (10)	
14.	Lub. Oil Pipe Lub. oil pipe (cylinder ~ head) Swivel pipe joint bolt 8 Packing (round copper) 14	1 3 6	Fix the lub. oil pipe (cylinder ~ head).		
15.	Valve Push Rod Cover Valve push rod cover Spring for valve push O-ring 16 O-ring 18	6 6 6 6	Assemble springs, the O-rings, and valve push rod cover. Then fix the assembly to the cylinder. Be sure to use right O-rings: Cylinder side ··· O-ring 16 Bonnet side ··· O-ring 18		

	Part Name	Piece	Order and Caution	Tool Used	Figure
16.	Bonnet				
	Bonnet ass'y	3	Attach bonnet fixing	Spanner	
	Packing	3	packings.		
	Small hex. nut	6	Confirm whether the valve		
	Valve push rod	6	push rod covers are firmly placed in the holes and O-rings are properly attached. Use small hex. nuts for tightening. Insert the valve push rods.		
	Valve rocker arm support ass'y	3	Fix the valve rocker arm support, making the centers of valve rocker arm concentric.	Spanner (21)	
	Hex. nut 14	3			
			After confirming that the pins are placed in the positioning holes of valve rocker arm, tighten with the nuts.		
			Confirm, by moving the valve push rods up and down, that there is cushion on push rod cover.		
17.	Lub. Oil Overflow Pipe (head ~ cylinder)	1	Attach the rub. oil overflow pipe (head ~ cylinder)	Spanner (19) Spanner (21)	
	Swivel pipe joint bolt 8	3			
	Packing (round copper) 14	6			
	Swivel pipe joint bolt 10	1			
	Packing (round copper) 16	2			

	Part Name	Piece	Order and Caution	Tool Used	Figure
18.	Exhaust Manifold Ass'y				
	Exhaust manifold ass'y	1	Attach the exhaust manifold.	Spanner (17)	
	Gasket packing	3	(Note) *Use stainless bolts.*		
	Hex. bolt (stainless) 10 × 25	6	Tighten evenly.		
	Hex. bolt (stainless) 10 × 32	3			
	Cooling water pipe (head ~ exh. manifold)	3	Attach the cooling water pipe.	Box spanner (14) Spanner (17)	
	Packing	6	(Note) *Do not attach*		
	Hex. bolt (plated) 8 × 45	12	*plug to the No. 3 cylinder. Attach by-pass pipe of*		
	Thin plug (brass) M12	2	*bilge pump.*		
	Packing (round copper) 12	2	*Use plated bolts.*		
19.	Lub. Oil Cooler Ass'y				
	Lub. oil cooler ass'y	1	Fix cooler, applying	Spanner (17)	
	Body packing (A)	1	Three-bond on the fixing	Box spanner (17)	
	Body packing (B)	1	side surface of cylinder.		
	Hex. bolt M10 × 25	3	Tighten the upper side first and then the lower side.		
	Hex. bolt M10 × 50	3	Connect cooling water pipe (water pump ~ cooler).		
	Lub. oil pipe (cooler outlet ~ cylinder inlet)	1	Attach lub. oil pipe (cooler outlet ~ cylinder inlet).	Double offset wrench (14)	
	Packing	2			
	Hex. bolt M8 × 20	1			
	Hex. bolt M8 × 32	1			
	Hex. bolt M8 × 36	2	Connect lub. oil pipe (cylinder ~ head) with lub. oil pipe (cooler outlet ~ cyl. inlet).	Double offset wrench (17)	
	Swivel pipe joint bolt 6	1			
	Packing (round copper) 12	2			

	Part Name	Piece	Order and Caution	Tool Used	Figure
20.	Breather Body Breather body Packing Hex. bolt M10 × 25 Hex. bolt M10 × 45	 1 1 1 1	Attach breather body.	Box spanner (17)	
21.	Starting Motor Starting motor Hex. bolt M12 × 50 Hex. nut M12 Spring washer 12	 1 3 3 3	Fix the starting motor.	Double offset wrench (19) Spanner (19)	
22.	Centrifugal Strainer Centrifugal strainer Centrifugal strainer support Packing Centrifugal strainer packing Hex. bolt M8 × 28 Hex. nut M8 Hex. bolt M10 × 25	 1 1 1 1 2 2 2	Assemble centrifugal strainer, centrifugal strainer support, and lub. oil pipe (cyl. outlet ~ centrifugal strainer) and then attach the assembly to the cylinder side cover.	Spanner (13) Spanner (17)	
	Lub. oil pipe (cyl. outlet ~ centrifugal strainer) Swivel pipe joint bolt 8 Packing (round copper) 14	1 1 2	Fix the lub. oil pipe (cyl. outlet ~ strainer inlet).		
	Lub. oil pipe (cyl. outlet ~ strainer inlet) Packing Hex. bolt M10 × 25 Hex. bolt M10 × 36	1 1 1 1	Connect lub. oil pipe (cyl. outlet ~ centrifugal strainer) with lub. oil pipe (cyl. outlet ~ strainer inlet).		

	Part Name	Piece	Order and Caution	Tool Used	Figure
23.	Charging Dynamo		Assemble charging dynamo and bracket. Attach to the stud bolt. (Note) *Tighten by using small hex. bolts.*	Spanner (17) Spanner (14) Box spanner (17)	
	Charging dynamo	1			
	Charging dynamo bracket	1			
	Hex. nut M10	4			
	Small hex. bolt M10 X 40	2			
	Spring washer	2			
	Hex. lock nut M10	2			
	V-belt adjuster	1	Attach the V-belt adjuster. Tighten bolts while lifting up the dynamo and stretching the belt.	Double offset wrench (14 X 17) Spanner (14 x 17)	
	Adjuster washer	1			
	Spring washer 8	1			
	Hex. bolt M8 X 22	1			
	Hex. bolt M10 X 20	1			
24.	Thrust Shaft		Attach the master gear to the thrust shaft and tighten with tightening nut.	Hammer Minus screwdriver Special box spanner	
	Thrust shaft	1			
	Reduction master gear	1			
	Single-row & deep-grooved ball bearing	1			
	Spacer	1			
	Thrust bearing washer	1			
	Shake proof washer	1			
	Tightening nut	1			
25.	Intermediate Gear		Hammer the intermediate gear shaft in. (Note) *Drive in, setting the oil hole up.*	Wooden hammer	
	Intermediate gear	1			
	Intermediate gear	1	Tighten with the hex. bolt to fix the intermediate gear. Bend the bend washer firmly.	Spanner (14)	
	Intermediate gear washer	1			
	Bend washer	1			
	Hex. bolt M10	1			

	Part Name	Piece	Order and Caution	Tool Used	Figure
	O-ring Intermediate gear shaft washer Shake proof washer Tightening nut	1 1 1 1	Assemble in the order of O-ring, intermediate gear shaft washer, shake proof washer, and tightening nut.	Hammer Minus screwdriver	
26.	Reduction Reversing Gear Lower Case Reduction reversing gear lower-case Packing for reduction reversing gear case Hex. bolt M10 X 25 Straight pin 10φ X 22	1 1 11 2	Stick the packing using 3-Bond on the attaching surface of cylinder. Attach the reduction reversing gear lower case.	Spanner (17) Double offset wrench (17)	
27.	Flywheel Flywheel Feather key (with screw) 15 X 65 End-nut Lock washer	1 1 1 1	Fix the flywheel. Tighten firmly by using end-nut spanner. (Note) *Use new lock washer.* *Tighten torque must be 40 kg-m.*	End-nut spanner Hammer	
28.	Clutch Housing Ass'y Clutch housing Flanged straight pin Lock bolt (housing ~ crankshaft coupling) Wire	1 3 6	Apply grease to the cylindrical roller bearing. Drive flanged straight pin in housing (A). Attach housing ass'y. Tighten housing (A) and rank shaft coupling by using tightening bolt. Bind tightening bolt by wire. (Note) *Pay attention not to drop the cut waste of wire into the reversing gear lower case.*	Copper hammer Double offset wrench (17) Spanner (17) Pliers	
29.	Governor Weight Governor weight Pin for governor Washer Cotter pin 3 X 25 Slide barrel Ball bearing #6202 Ball bearing caulking piece	2 2 2 2 1 1 1	Attach the governor weight and then lock by using cotter pin. Confirm whether the governor weight works lightly.	Radio pliers	

	Part Name	Piece	Order and Caution	Tool Used	Figure
	Governor Weight		Assemble slide barrel, ball bearing and caulking piece. Then attach the assembly. Confirm whether the slide barrel is sliding lightly.		
30.	Bilge Pump				
	Bilge plunger	1	Connect the plunger to double offset link, and lock by using shaft lock ring.	Radio Pliers	
	Plunger pin	1			
	Shaft lock ring 10	2			
	Cylinder gear chamber side cover ass'y	1	Stick the packing using 3-bond adhesive on the attaching surface. Attach the cylinder gear chamber side cover ass'y.	Spanner (17)	
	Gear side cover packing	1			
	Hex. bolt M10 × 25	1			
	Hex. bolt M10 × 36	3			
	Bilge pump	1	Insert water drip rubber, oil seal and plunger guide. Tighten nut, attaching bilge pump,	Spanner (17)	
	Plunger guide	1			
	Plunger water drip rubber	1			
	Oil seal	1			
	Hex. nut M10	2			
	Pump packing	1			
	Cooling water by-pass pipe (for bilge pump)	1	Connect cooling water by-pass pipe with bilge pump and cooling water pipe.	Spanner (21)	
	Swivel pipe joint bolt	1			
	Packing (round copper 12)	2			
	Steady rest for 6φ pipe	1			
31.	Reduction Reversing Switch Gear Upper Case & Switch Valve Ass'y				
	Reduction reversing gear upper case	1	Temporalily attach the reduction reversing upper case.		
	Hex. bolt M10 × 25	17			
	Hex. bolt M12 × 71	2			

	Part Name	Piece	Order and Caution	Tool Used	Figure
	Switch valve ass'y Packing Hex. bolt M10 × 25	1 1 4	Attach the switch valve ass'y.	Spanner (17)	
	Straight pin with 8φ screw Hex. nut M8	2 2	Drive the straight pin in and tighten reduction reversing gear upper case by using bolt.	Box spanner (14 × 17)	
32.	Fuel & Lub. Oil Strainer Ass'y Fuel lub. oil strainer ass'y Hex. bolt M10 × 148 Hex. bolt M10 × 100 Fuel oil pipe union Packing (round copper) 18	1 1 2 1 1	Attach the fuel & lub. oil strainer ass'y.	Box spanner (17)	
	Fuel oil pipe (strainer ~ pump) Pipe joint bolt with air hole Air vent plug Packing (round copper) 16 Packing (round copper) 8	1 4 4 8 4	Attach the fuel oil pipe (strainer ~ pump) to strainer and pump.	Double offset wrench (21)	
33.	Lub. Hydraulic Oil Pipe & Oil Pressure Adjusting Valve Lub. oil pipe (cyl. outlet ~ strainer inlet) Packing Hex. bolt M8 × 20	 1 2	Connect the lub. oil pipe (cyl. outlet ~ strainer inlet) with strainer inlet.	Double offset wrench (14)	

Part Name	Piece	Order and Caution	Tool Used	Figure
Lub. oil pipe (strainer outlet ~ cooler inlet)	1	Attach lub. oil pipe (strainer outlet ~ cooler inlet).	Double offset wrench (14)	
Packing	2			
Hex. bolt M8 × 20	2			
Hex. bolt M8 × 32	2			
Oil pressure adjusting valve ass'y	1	Attach oil pressure adjusting valve ass'y.	Double offset wrench (17)	
Packing	1			
Hex. bolt M10 × 22	2			
Hydraulic oil strainer	1	Attach hydraulic oil strainer and working oil adjusting valve ass'y.	Double offset wrench (17)	
Hydraulic oil adjusting valve ass'y	1			
Packing	1			
Hex. bolt M10 × 71	1			
Hex. bolt M10 × 90	1			
Hydraulic oil pump (cyl. outlet ~ hydraulic oil pump)	1	Attach hydraulic oil pipe (cyl. outlet ~ working oil pump).	Spanner (26)	
Packing (round copper) 20	2			
Swivel pipe joint bolt	1			
Packing	1			
Hex. bolt M10 × 28	1			
Hex. bolt M10 × 40	1			
Hydraulic oil pump (pump ~ strainer)	1	Connect hydraulic oil pipe (pump ~ strainer) with hydraulic oil pump and hydraulic oil strainer.	Spanner (26) Spanner (32)	
Swivel pipe joint bolt 15	1			
Packing (round copper) 20	2			

	Part Name	Piece	Order and Caution	Tool Used	Figure
	Hydraulic oil pipe (strainer ~ switch valve)	1	Connect hydraulic oil pipe (strainer ~ switch valve) with hydraulic oil strainer and switch valve.	Spanner (21) Spanner (23)	
	Pipe joint bolt	1			
	Packing (round copper) 18	2			
	Swivel pipe joint bolt	2			
	Packing (round copper) 16	2			
	Hydraulic oil pipe (rear cover ~ intermediate shaft)	1	Connect hydraulic oil pipe (rear cover ~ intermediate shaft) with rear cover and intermediate shaft.	Double offset wrench (21)	
	Swivel pipe joint bolt	2			
	Packing (round copper) 12	4			
34.	Gauge Board Ass'y & Engine Lift Metal Fitting		Assemble the gauge board ass'y and the engine lifting metal and attach to the cylinder.	Double offset wrench (21)	
	Gauge board ass'y	1			
	Engine lifting metal	1			
	Hex. bolt M12 X 25	2			
	Spring washer	2			
	Hydraulic oil pipe (strainer ~ pressure gauge)	1	Connect hydraulic oil pipe (strainer ~ pressure gauge) with hydraulic oil strainer and pressure gauge.	Spanner (17)	
	Packing (round nylon) 5	1			
	Lub. oil pipe (cylinder ~ pressure gauge)	1	Connect lub. oil pipe (cylinder ~ pressure gauge) with cylinder and lub. oil pressure gauge.	Spanner (17)	
	Swivel pipe joint bolt 6	1			
	Packing (round copper) 12	2			
	Packing (round nylon) 5	1			

Part Name	Piece	Order and Caution	Tool Used	Figure
35. Pre-Combustion Chamber & Fuel Injection Valve Pre-combustion chamber (front chamber) Pre-combustion chamber (rear chamber) Packing (front chamber) Packing (rear chamber)	3 3 3 3 3	Set the pre-combustion chamber in the order of packing (for front chamber), pre-combustion chamber (front chamber), packing (for rear chamber), and pre-combustion chamber (rear chamber). (Note) *Be sure to use proper packings.* Packing (front chamber) ..... bore 29φ Packing (rear chamber) ..... bore 35.5φ		
Fuel injection valve Fuel valve holder Hex. nut 10	3 3 6	Attach the fuel injection valve and lock with fuel valve holder and tighten with nut. (Note) *Tighten the fuel valve holder evenly.*	Spanner (17)	
High pressure fuel pipe	3	Connect the high pressure fuel pipe with fuel pump and fuel injection valve.	Spanner (17)	
Fuel overflow pipe (fuel valve side) Swivel pipe joint bolt 6 Packing (round copper) 12	1 3 6	Attach the fuel overflow pipe (fuel valve side).	Spanner (17)	
36. Governor Housing Governor housing ass'y Governor spring Governor spring for low speed Spring holder Packing Hex. bolt M8 × 40 Hex. bolt M8 × 50	1 1 1 1 1 2	Insert the governor housing ass'y while pressing the spring by hand in order not to detach the governor spring from the spring. Confirm that the governor lever shaft works lightly.	Spanner (14)	

	Part Name	Piece	Order and Caution	Tool Used	Figure
	Governor link	1	Connect the governor lever and the fuel pump by using governor link.	Special spanner (10) Pliers Minus screwdriver	
	Pin	1			
	Washer	1			
	Cotter pin 1.6φ×10	1			
	Lock nut M5 (left screw)	1			
	Lock nut M5 (right screw)	1			
37.	Regulator Handle Support Ass'y				
	Regulator handle support ass'y	1	Attach the regulator handle support.	Spanner (14)	
	Hex. bolt M8 × 20	2			
	Governor link	1	Connect the governor link with the regulator handle lower part and the regulator lever.	Special spanner (10)	
	Joint (A)	1			
	Joint (B)	1			
38.	Tachometer Flexible Shaft				
	Tachometer flexible shaft	1	Attach the tachometer flexible shaft.	Spanner (35)	
39.	Suction Port Cover & Bonnet Cover				
	Suction port cover	1	Attach the suction port cover.	Box spanner (14)	
	Hex. bolt 8 × 50	2			
	Bonnet cover	3			

6. DISASSEMBLY AND REASSEMBLY OF OTHER PARTS

6-1. Fuel Injection Pump

When the fuel injection pump is to be disassembled or reassembled, handle the parts with extra care not to leave any dust or to incur scratch to the parts.

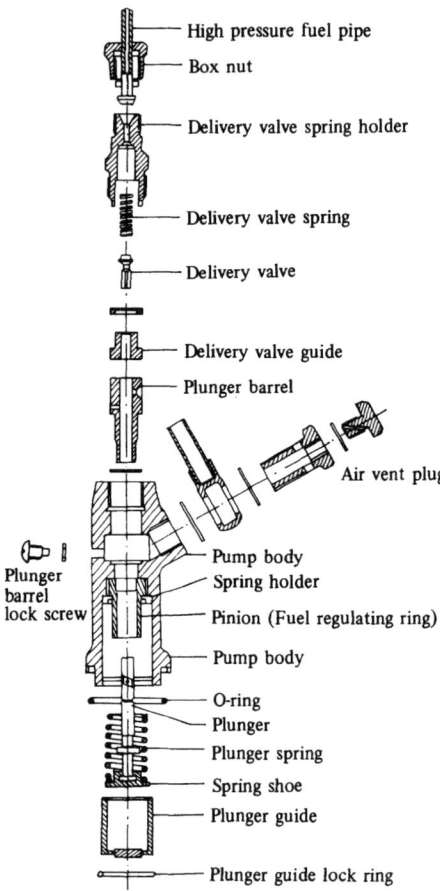

Cross-Section View of Fuel Injection Pump

(1) Disassembling Procedure

	Procedure and Caution	Tool Used	Figure
1	1) Loosen the box nut of high pressure fuel pipe and remove it.	Spanner (19)	
	2) Loosen the bolt of fuel oil pipe and remove it.	Spanner (17)	
	3) Remove the fuel regulating rack adjusting shaft.	Spanner (10)	
	4) Remove the fuel pump seat bolt.		
	5) Remove the fuel pump.		
2	1) Remove the delivery valve spring holder.	Spanner (21)	

	Procedure and Caution	Tool Used	Figure
	2) Remove the spring for delivery valve guide and the delivery valve spring seat.		
	3) Pull out the fuel injection valve guide. Confirm if both the delivery valve and delivery valve guide are free of dust, scratch or worn-out.	Delivery valve-guide removing tool Spanner (26)	
3	1) Remove the plunger guide lock ring.	Pliers	
	2) Loosen the plunger barrel lock bolt.	Spanner (10)	
	3) Remove plunger guide, plunger, plunger spring, plunger spring shoe, plunger spring holder and plunger barrel. Remember the disassembling order of each part. Wash each part thoroughly. Check whether the plunger is free of dust, scratch, and its tip is not discolored or worn. If any abnormality is found at the tip of plunger, exchange both the plunger and the plunger barrel.		

(2) Assembling Procedure

	Procedure and Caution	Tool Used	Figure
1	1) Insert the plunger barrel and set. Be sure to adjust the direction of the plunger barrel in order to insert lock bolt.		
2	1) Place the fuel pump body upside down.		
	2) Insert the fuel regulating ring. Pay attention to match cut mark of fuel regulator rack with the punch mark of fuel regulator shaft. (Refer to the right figures.)		Cut mark / Punch mark
	3) Insert the plunger. The direction of the plunger is attained by matching the 0-point mark on the plunger side and the cut mark on the fuel regulator ring side.		0-point mark / Cut mark
	4) Insert the plunger spring holder, plunger spring, the plunger spring shoe, the plunger guide, and set the plunger guide lock ring.	Pliers	
3	1) Place the fuel pump body upright.		

— 51 —

Procedure and Caution	Tool Used	Figure
2) After inserting the fuel delivery valve into fuel delivery valve guide, insert them into fuel pump body.		
3) Insert delivery valve spring holder washer and spring for fuel delivery valve guide.		
4) Reassemble delivery valve spring holder.		

6-2. Fuel Injection Valve

The fuel injection valve of Model SM engine is designed to atomize the fuel oil delivered from the fuel pump under high pressure so that the fuel oil may be combusted easily. If the atomization takes place incompletely, a major cause for inferior engine performance will occur.

There are a nozzle valve in order to complete atomization and a pressure regulating device to make injection of a fuel oil possible under high pressure. Since precise processing was given to the nozzle for this purpose, handle the nozzle with extra care not to incur any scratch or to leave dust.

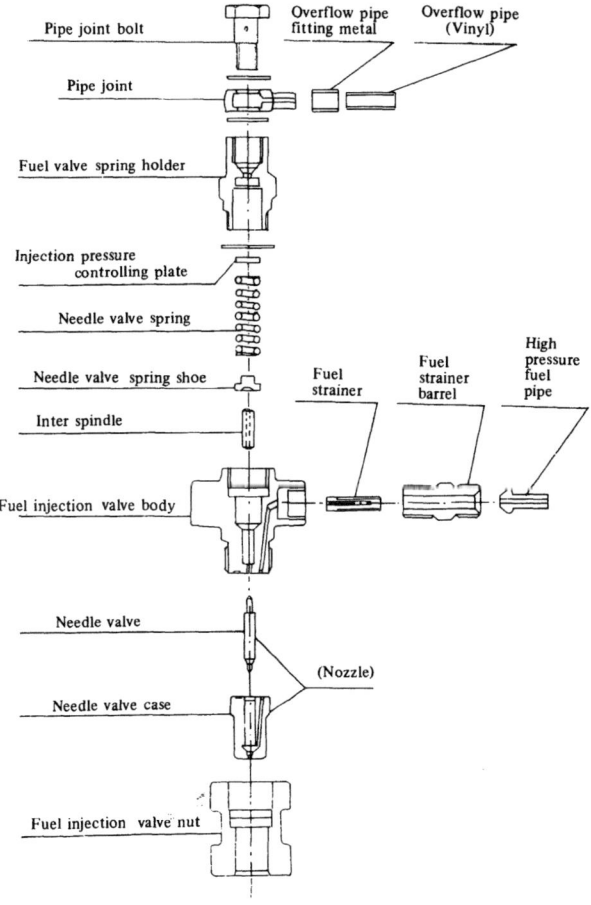

Disassembled, Cross-Section
View of Fuel Injection Valve

(1) Disassembly of Fuel Injection Valve

	Procedure & Caution	Tool Used	Figure
1	1) Loose the lock nut of fuel injection valve holder and then remove the fuel injection valve holder. 2) Pull out the fuel injection valve. (Note) *Remember the position of the arrow mark on the injection pipe side of the fuel injection valve.*		
2	1) Remove the fuel injection valve nut. 2) Detach the needle valve case and needle valve from the fuel injection valve nut. (Note) *When it is difficult to remove the nozzle, use a nozzle-body removing tool and remove it in the manner of right figure.* *Pull out the nozzle valve and check if the seat part of needle valve is free of dust and scratch or the sliding part is not discolored or worn off. If there is no abnormality, wash the needle valve and place it in the needle valve case.*	Spanner (26) Nozzle removing tool	
3	1) Remove the fuel valve spring holder.	Spanner (19)	

	Procedure and Caution	Tool Used	Figure
	2) Detach the needle valve spring, needle valve spring shoe and inter spindle. (Note) *Pay attention not to lose the injection pressure regulating plate.*		
4	1) Remove the fuel strainer barrel.	Spanner (14)	
	2) Remove the fuel strainer. (Note) *If it is difficult to remove the fuel strainer, use a thinner nozzle removing tool.*	Nozzle removing tool	

(2) Reassembling Fuel Injection Valve

	Procedure and Caution	Tool Used	Figure
1	1) Insert the needle valve and needle valve case into the fuel injection valve nut. (Note) *Wash them with clean oil and smear oil.*		
	2) Set the fuel injection valve nut to the fuel injection valve body.	Spanner (26)	
	3) Insert the interspindle, needle valve spring shoe, needle valve spring, and injection pressure regulating plate into the fuel injection valve body.		

	Procedure and Caution	Tool Used	Figure
	4) Attach the fuel valve spring holder.	Spanner (19)	
	5) Insert the fuel strainer into the fuel strainer barrel.		
	6) Attach the fuel strainer barrel to the fuel injection valve body.	Spanner (14)	

6-3. Cylinder Head

(1) Disassembly of Cylinder Head

	Procedure and Caution	Tool Used	Figure
1	1) Remove the cylinder head from cylinder body.		
2	1) Remove the anti-chamber.		
3	1) Attach the hex. nut (14mm) to valve lever supporting bolt.	Suc./Exh. valve handling tool	

Procedure and Caution	Tool Used	Figure
2) Remove the valve stud by using the suc./exh. handling tool in the manner of right figure. 3) Remove the suc./exh. valve.		

(2) Reassembly of Cylinder Head

	Procedure and Caution	Tool Used	Figure
1	1) Insert the suc./exh. valve. (Note) *Do not misplace the suc. and exh. valves.*		
	2) Set the valve spring, inside spring and valve spring shoe.		
	3) Attach the valve stud by using the suc./exh. valve handling tool.	Suc./exh. handling tool	
2	1) Reassemble in the order of the packing for pre-combustion chamber (front chamber), pre-combustion chamber (front chamber), the packing for pre-combustion chamber (rear chamber), and pre-combustion chamber (rear chamber). (Note) *Do not place the pre-combustion chamber (rear chamber) upside down.*		

6-4. Replacement of Cylinder Liner

Cylinder liner is plated with hard chromium so that the serviceable period can be prolonged to considerable extent. However, improper handling and maintenance may shorten the period. In case of seizure or worn-out, cylinder liner should be replaced with a new one. This replacement must be carried out properly; otherwise the seizure, or water and oil leakage may take place again within a short period. Therefore it is advisable to understand thoroughly following replacement procedure.

(1) Procedure of Removing Cylinder Liner

Procedure and Caution	Tool Used	Figure
1) Attach the lower liner removing plate to the cylinder liner outskirts.	Liner removing tool Spanner	
2) Insert the cylinder liner removing bolt into the cylinder liner skirt. (Note) *Be sure to fit nuts to the liner removing bolts.*		
3) Fit the cylinder head tightening nuts to two diagonal cylinder head set bolts.		
4) Fit upper liner removing plate to the cylinder liner removing bolts and cylinder head bolts.		
5) Attach and tighten liner removing bolt nuts. (Note) *Tighten the nuts until the cylinder liner can be removed.*		

(2) Procedure on Inserting Cylinder Liner

Procedure and Caution	Tool Used	Figure
1) Remove the liner rubber packing, and clean the cylinder liner inserting part of the cylinder body. (Note) *Since the cylinder liner inserting part is coated with a white paint, scrape off the paint by using sandpaper of #80~120.*		
2) Wash new cylinder liner to remove rustproof oil.		
3) Paint outer cylinder liner inserting part into white.		
4) Attach cylinder liner rubber packing. (Note) *Always use new rubber packing for cylinder liner.* *Do not twist the packing.*		
5) Insert the cylinder liner.		

6-5. Piston Rings, Piston and Connecting Rod

Piston rings fitted to piston are made of special cast iron in order to withstand high pressure and high temperature. However, to prevent seizure, sticking, and oil pumping up, and to ensure its durability, rings of various shapes are being used. Therefore, at the time of changing piston rings, be certain to use proper kinds of YANMAR parts in proper arrangements.

(1) Disassembly and Reassembly of Piston Rings

When engine has been operated for long period or handled carelessly, seizure, sticking, or abrasion of piston rings will occur. These may become a cause of compression leakage. In such cases, exchange the piston rings in the following manner:

	Procedure and Caution	Tool Used	Figure
1	1) Remove piston from the engine, and inspect the condition of piston rings. *Refer to Page 89 ~ 97 for determining unfitness and abrasion limit.*		
2	1) Make 2 wire rings of about 5cm diameter.	Wire	
	2) Hook wire rings to piston ring cut, and put thumbs through the rings.		
	3) Apply a strength on thumb, and, expanding the piston ring, carefully pull up the piston ring by middle fingers. (Note) *Take care not to expand the piston ring too much, otherwise it might be broken.*		

	Procedure and Caution	Tool Used	Figure
3	1) Clean the piston ring groove sufficiently by removing the carbon deposited. (Note) *Unless carbon sticking to the ring groove is removed, the ring cut will not fit, and the piston will not properly be into the cylinder liner.*	Minus screwdriver and damaged ring	
4	1) Be sure to make distinction among the piston rings. (Note) *1st, 2nd & 3rd rings ····· compression rings* *4th & 5th rings ········ oil rings*		

(2) Disassembly and Reassembly of Piston and Connecting Rod

Piston and connecting rod are connected by means of piston pin which is sliding inside the smaller end bush of connecting rod.

In operation, piston pin inserting hole expands due to high temperatures and the clearance to piston pin becomes larger; therefore at normal temperature, some margin is given thereto. The piston pin can be easily inserted or pulled out by heating the piston up to approx. 80°C. The removing and inserting procedure are as follows:

1) Removing Procedure of Piston and Piston Pin

	Procedure and Caution	Tool Used	Figure
1	1) Remove the piston from the engine, and inspect the condition of piston. (Note) *Refer to 7-3 for determing unfitness and abrasion limit.*		
2	1) Remove piston pin circlips.	Pliers	
3	1) Warm oil (may use hot water). (Note) *In case of oil, use light oil, heavy oil, or engine oil and warm it to approx. 80°C. Care should be paid to prevent oil from ignition.*		

	Procedure and Caution	Tool Used	Figure
	2) Warm the piston. (Note) *Immerse in oil of around 80°C for 10~15 minutes.*		
	3) Take out the piston from oil and by using the piston pin removing tool, lightly hammer the piston pin. (Note) *Swiftly hammer the piston pin out of the piston while it is still warm.*	Piston pin removing tool Hammer	

2) Procedure on Reassembling Piston and Piston Pin

		Procedure and Caution	Tool Used	Figure
1	1)	Set the piston pin circlip only for one side.	Pliers	
	2)	Warm the piston. (Note) *Immerse in oil of about 80°C for 10~15 minutes. May use hot water.*		
2	1)	Take out the piston and place it (being upside down) on repairing table.		
	2)	Insert the connecting rod into the piston. Then insert piston pin by turning the piston pin from the side of no piston pin circlip set. (Note) *It is better to have oil applied on piston pin.* *Swiftly insert them while the piston is still warm.* *Check if the connecting rod and piston pin move lightly.*		

3) Replacement of Small End Bush of Connecting Rod

	Procedure and Caution	Tool Used	Figure
1	1) Place the small end part of connecting rod on repairing table (vise stand) so that the bush is easily removed.		
	2) Take out the small end bush by tapping with small end bush removing tool.	Small end bush removing tool	
2	1) Prepare new small end bush.		
	2) Fit the bush, matching the oil hole of new small end bush with the oil hole of small end part of connecting rod.		
	3) Insert the new small end bush by using the small end bush inserting tool. (Note) *Instead of tapping with a hammer, it is better to press the bush in by a vise or press.*	Small end bush inserting tool	
	4) Insert the piston pin into small end bush, and check if it slides lightly.		

6-6. Clutch Housing

1) Disassembly of Clutch Housing

Part Name	Piece	Disassembly Procedure	Tool Used	Figure
Disassembly of Clutch Housing Hex. bolt (for housing (A)-(B)-(C))	12	1) Cut the wire and remove the hex. bolts.	Pliers Spanner (14) Double offset wrench (14)	
Housing (C)	1	2) Insert 8mm-bolt into dismantling hole (4 places) and tighten the bolts. Then housing (C) can be removed.		
Ahead shaft bearing tightening nut	1	3) Remove it.	Special spanner Hammer	
Cylindrical roller bearing (for ahead shaft)	1	4) Remove the cylindrical roller bearing (for ahead shaft). (Note) *Remember the direction of the bearing.*		
Reduction pinion	1	5) Pull it out by using reduction pinion removing tool.	Reduction pinion removing tool Spanner	
Feather Key	1	6) Pull out the feather key using M5 bolt.		

Part Name	Piece	Disassembly Procedure	Tool Used	Figure
Single-row, deep-grooved ball bearing Astern friction plate } ass'y Astern shaft Bearing holder spring	1 1 1	7) Remove the single-row, deep-grooved ball bearing, friction plate, astern shaft ass'y, and bearing holder spring.		
Hex. socket bolt (for oil pressure hydraulic barrel)	6	8) Remove the hex. socket bolt (for oil pressure hydraulic barrel).	8mm-hex. wrench	
Housing (A)(B)	1	9) Separate the housing (A) from housing (B).		
Oil pressure hydraulic barrel (B)	1	10) Screw in the 8mm-bolt into the dismantling hole (3 places) and tightening them, then the oil pressure hydraulic barrel (B) can be detached from the housing (B).		

2) Reassembly of Clutch Housing

Part Name	Piece	Reassembly Procedure	Tool Used	Figure
Housing (B) O-ring for housing (B) Oil pressure hydraulic barrel (B) O-ring for oil pressure hydraulic barrel (B)	1 1 1 2	1) Attach the oil pressure hydraulic barrel (B) to the housing (B). (Note) Use new O-ring.	Wooden hammer	
Oil pressure hydraulic barrel (A) Neutral retaining spring Straight pin (A) (for oil pressure hydraulic barrel) Straight pin (B) (for oil pressure hydraulic barrel)	1 4 4 3	2) Attach the neutral retaining spring and straight pins (A)(B) to the housing (B).		

Part Name	Piece	Reassembly Procedure	Tool Used	Figure
Hex. socket bolt (for oil pressure hydraulic barrel)	6	3) Tighten using the hex. socket bolt (for oil pressure hydraulic barrel).	8mm-hex. wrench	
		4) Attach wire.	Wire	
Ahead shaft	1	5) Attach ahead shaft and friction plate for ahead. (Note) *Apply sufficient oil to the metal part.*		
Friction plate for ahead	1			
Housing (A)	1	6) Attach the housing (A). (Note) *Apply 3-bond evenly.*	Wooden hammer	
Single-row, deep-grooved ball bearing	2	7) Attach lach part according to the order shown in the right figure.		
Astern friction plate	1			
Astern shaft	1			
Bearing holder spring	1			
Reduction gear	1			
Cylindrical roller bearing	1			
Shake proof washer	1			
Ahead bearing tightening nut	1			
Feather key				
Housing (C)	1	8) Evenly apply 3-bond on housing (C) and tighten using hex. bolt.	Spanner (14)	
Hex. bolt (for housing (A)(B)(C))	12			
Emergency bolt wire	4	9) Attach the emergency bolt.		

7. ADJUSTMENT AND SERVICING

7-1. How to Adjust the Adjusting Part

(1) Adjustment of suction & exhaust valve head clearance

If the clearance between the suction/exhaust valve and the valve lever is too wide, the timing valve operation will change, and this may result in decreased horsepower, defective exhaustion or difficult starting.

If there is no clearance, compression leakage will take place from valve seats. The periodical inspection and adjustment are therefore necessary. (This should be done after initial 50 hours and thereafter every 300 hours.)

The adjustment of clearance must be performed when the engine is cool. Stop the engine at top dead center of compression and adjust the clearance between the valve and valve lever to 0.1mm for both suction and exhaust valves by the adjusting screws.

Adjusting procedure

Procedure	Instruction	Tool Used	Figure
Preparation 1. Adjust the No.1 cylinder first. Turn the flywheel and stop at its compression top dead center (1T mark)	1) Decompress No.2 and No.3 cylinders. 2) Turn the flywheel direction of revolution. 3) Set the arrow to 1T mark.		
2. Loosen the adjusting screw nut of valve clearance.		Spanner (17)	
Adjustment 3. By the valve clearance adjusting screw, adjust the clearance to 0.15mm.	Adjust it to such an extent that the gauge can move lightly.	Thickness gauge	
4. Fix the adjusting screw by the valve clearance adjusting screw nut.	When tightening the lock nut, hold the adjusting screw by a minus screwdriver, then tighten it by spanner.	Minus screwdriver Spanner (17)	
5. Check the clearance after adjustment.	Check if the clearance is properly adjusted.	Thickness gauge	
6. Similarly adjust No.2 and No.3 cylinders.			

— 67 —

(2) Adjustment of Governor Link

The function of governor system corresponds to that of the nervous system of human body; that is, it accurately detects the variation of load placed on engine and to control the amount of fuel to be sent to the fuel injection pump in order to ensure stable rotation. Namely, if change in revolution occurs due to load, the governor weight is actuated to transmit this change to the governor weight is actuated to transmit this change to the governor link, and furthermore the quantity of fuel oil is regulated as the mechanism is interlocked with the fuel pump.

Therefore, if the adjustment of governor link is incorrect, the action of governor weight will not accurately be transmitted to the fuel injection pump, and such way cause irregular revolution, decreased horsepower, or defective starting.

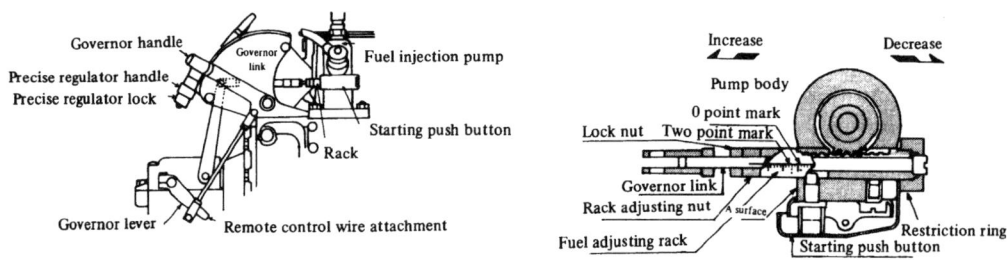

Adjusting procedure (governor link)

Procedure	Instruction	Tool Used	Figure
1. Set the regulator handle at stop position.			
2. Set the rack to two-point mark.	Set all three cylinders to two-point mark.	Spanner (10)	
3. Set the regulator handle at maximum position.			

Procedure	Instruction	Tool Used	Figure
4. Push the injection regulating fittings (rack push button).			
5. Adjust No. 2 governor link.	Turn the governor link and adjust the clearance between the rack and the lock nut to 0.1mm.	Spanner (10)	
6. Check the fuel injection timing. (Note) *As for the adjustment of fuel injection timing, refer to page 76.*			
7. Start the engine.			
8. Set the regulator handle at maximum position.			
9. Adjust the governor link so that the rotation of engine becomes 2100 rpm.	The rpm is decreased by expanding No. 1 governor link and increased by shortening it.	Spanner (10)	

(3) Adjustment of Fuel Injection Pressure
The injection pressure of the fuel injection valve has a great influence over the combustion at explosion. If the pressure is inappropriate, it will cause not only decreased horsepower, difficult starting, uneven revolution or other improper operation but also sticking or early wear of nozzle and carbon deposite inside the combustion chamber and other troubles.

Adjust the pressure by using a nozzle tester so that the injection pressure becomes 160 kg/cm^2.

If the fuel is not injected properly even after completion of the adjustment, perform fitting for the nozzle (needle valve) or change it with new one.

Procedure	Instruction	Tool Used	Figure
Preparation 1. Remove the high pressure fuel pipe.		Spanner (19)	
2. Detach the fuel injection valve ass'y.	Remove the fuel valve holder and detach the fuel injection valve ass'y.	Spanner (17)	
3. Attach the nozzle tester to the delivery valve and discharge air.	1) Attach the tester as shown in the figure. 2) Set the regulator handle at operation position. 3) Operate the priming lever until fuel comes out from the tester high pressure pipe.	Spanner (17)	

Procedure	Instruction	Tool Used	Figure
4. Attach the fuel injection valve ass'y to the tester high pressure pipe (outlet side).			
Measuring 5. Measuring of fuel injection pressure.	1) While operating the priming lever, read out the maximum indication at the time of fuel injection. 2) Check if oil is properly injected. (Note) *In case of improper injection, refer to 89 ~ 94. (Trouble And Countermeasures).*		15° Good Bad
Adjustment (In case of improper injection pressure) 6. Remove the fuel valve spring holder.	(See the description for fuel injection valve.)	Spanner (26) Spanner (19)	

7. Regulate the spring adjusting plate.	Proper adjustment will be attained by combining the plates with different thickness as follows:
	Thickness of plate Pressure
	    0.1 mm 7 kg/cm^2
	    0.2 ″ 14 ″
	    0.3 ″ 21 ″
	    0.4 ″ 28 ″
	    0.5 ″ 35 ″
	If the injection pressure is too high, remove certain number of the plates; if it is too low, add some.		

8. Tighten the fuel valve spring holder.		Spanner (19)	
9. Check the fuel injection pressure and atomization of fuel.			

(4) Adjustment of Fuel Injection Timing

Proper timing of injecting fuel from the valve is 8~12 degrees before the upper dead point. If the timing is too advanced or too lagged, it may cause some troubles such as difficult starting, knocking, insufficient horsepower, overheating and defective exhaust.

Before adjusting fuel injection timing, adjust the governor link and fuel injection pressure at the valve.

At the same time, fuel injection timing tends to delay due to wear of fuel pump plunger after a long operation of engine.

If the plunger is worn greatly, change the plunger, then check the fuel injection timing and adjust it.

Checking on Fuel Injection Timing

Procedure	Instruction	Tool Used	Figure
1. Remove the high pressure fuel pipe from fuel injection pump.		Spanner (17)	
2. Set the two-point mark of rack to surface (A) of the pump body.	Do not move the rack before completing the confirmation of all cylinder is completed.		
3. While watching the oil level of discharge hole of fuel injection pump delivery valve holder screw, turn the flywheel slowly in the direction of revolution.			
4. Stop the flywheel at the moment when oil level moved and read out the indication at the position.	Repeat 3 to 4 times the procedures in 3 and 4 and confirm the fuel injection timing.		

Procedure	Instruction	Tool Used	Figure
5. As graduations are provided on the flywheel every ten degrees before top head, it is proper if the graduation is indicated at 9 degree before top dead center. If the indication is improper, the injection timing should be adjusted.			

Adjustment of Fuel Injection Timing

Procedure	Instruction	Tool Used	Figure
1. Remove the adjusting window below the fuel injection pump.		Spanner (14)	
2. Loosen the loop nut and adjust the injection timing by adjusting bolt.	The fuel injection timing is delayed by screwing in the adjusting (counterclockwise) and quickened by screwing it out (clockwise). Set the spanner (10) on the adjusting bolt and turn it from the extreme right to the extreme left of the adjusting window, and one degree of variation will be gained.	Spanner (10) Spanner (17)	Adjusting bolt Lock nut
3. After adjustment, tighten the lock nut firmly so that the adjusting bolt will not become loose.	This work should be done carefully using two spanners so that the adjusting bolt will not turn with the lock nut.		
4. Reconfirm the fuel injection timing.			

(5) Adjustment of Lub. Oil and Hydraulic Oil Pressure

1) Adjustment of Lub. Oil Pressure

The engine side lub. oil pressure is set at $2 \sim 2.5$ kg/cm² at the time of delivery from factory. The pressure may sometimes become 4 kg/cm² immediately after starting the engine under cold weather conditions. This will not cause any trouble however.

The pressure should not be lower than 0.5 kg/cm² during low speed idle operation or lower than 2 kg/cm² during the operation at 1500 rpm.

If the pressure is not $2 \sim 4$ kg/cm² with warm oil, adjust the pressure according to the following procedures.

Procedure	Instruction	Tool Used	Figure
1. Remove the box nut of lub. oil pressure adjusting valve.		Spanner (19)	Cap / Lock nut
2. Loosen the lock nut and adjust the oil pressure by turning the spring holder.	Pressure is increased by turning the screw clockwise and is decreased by turning it counterclockwise.	Minus screwdriver	

2) Adjustment of Hydraulic Oil Pressure

Clutch side hydraulic oil pressure is proper if the pointer is indicating within the green range of $9 \sim 9.5$ kg/cm² at 1500 forward revolution.

If it is indicated outside the green range with warm oil, adjust the pressure by hydraulic oil pressure adjusting valve.

The adjusting procedure is the same as in the case of lub. oil pressure.

Hydraulic oil pressure is lowered by around 2 kg/cm² at the neutral position. This is due to the escape of hydraulic oil from switch valve, and such will not cause any trouble.

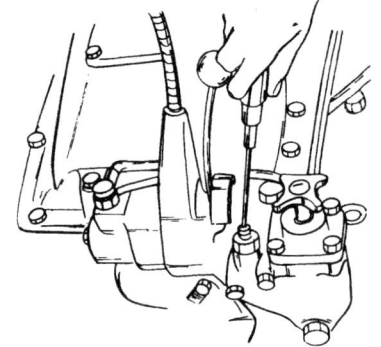

7-2. Servicing and Repairing of Major Parts

As the engine is operated for a long time, the operation may be interferred by such factors as wear of parts and others. The followings are major repairing and servicing procedures.

(1) General Cautions

Pay attention to the following cautions for maintenance and cleaning of engine:
1) Carefully observe the parts prior to repairing or cleaning for carbon deposit and contact. This practice will be very helpful for maintenance of engine.
2) Be careful not to scratch the parts when removing the carbon deposit.
3) Avoid using a wire brush or sandpaper for cleaning the precision parts, such as the contact or fitting surface.
4) Use a clean washing oil. Light oil is recommended.
5) Thoroughly clean the outside of engine block and the inside of crankcase at the time of disassembly. Also clean those parts and places which are difficult to clean after assembly.
6) Clean any rusty part with a fine sandpaper and apply oil afterward.
7) Remove scratches or burr on parts. It is not necessary to completely remove large ones but it is desirable to make the surface smooth.
8) If the metal fittings are accidentally hit with each other carefully make the surface smooth.

(2) Tightening of Cylinder Head

The cylinder head bolts are subjected to a large impact created at the time of explosion, which should therefore be tightened firmly, otherwise gas or water leakage may occur. If the stud bolts are loose, it is necessary to tighten them by using double nuts before fitting the cylinder head.

1) Tightening of Stud Bolts with Double Nuts

Procedure	Instruction	Tool Used	Figure
1. Screw in two nuts onto the bolt.	Put two nuts onto screw part opposite.		
2. Fix the nuts on the bolt.	Using 2 spanners prevent the inner nut from moving and tighten the outer nut firmly.		Tighten two nuts together
3. Set the stud bolt.	Apply the spanner to the outer nut, then tighten the stud bolt.		
4. Loosen the nuts from the bolt.	Lock the inner nut with spanner, preventing it from moving, and loosen the outer nuts, thus loosening both nuts.		

(3) Tightening of Connecting Rod Bolts
The crank pin metal is tightened by two connecting rod bolts. Since they receive a large force of inertia every revolution of crankshaft, insufficient clamping or bending of lock washers may cause serious accidents.
At the time of disassembling and servicing, tighten the two nuts uniformly and securely, and be certain to properly bend the lock washers.
Proper tightening torque of rod bolt is 9 kg-m (65 ft-lb).

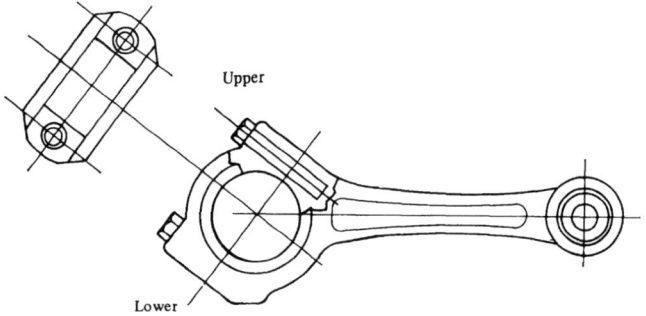

(4) Air Venting of Fuel Injection Pump
Air should be discharged because the engine will stop during operation if fuel runs out or after adjusting the fuel pump, the air may enter into the fuel pump to interrupt injection.

1) **How to Discharge Air**

Procedure	Instruction	Tool Used	Figure
1. Open the fuel cock.			
2. Loosen air venting small screw of fuel strainer.	Loosening small screw causes air mixed fuel to come out. Fuel should be injected until no foam comes out.	Spanner (14)	
3. Tighten air venting small screw.			

Procedure	Instruction	Tool Used	Figure
4. Loosen air venting small screw of fuel pump.	Injection should be continued until fuel without air mixed comes out.	Plus screwdriver	
5. Loosen air venting small screw.			
6. Loosen both ends of fuel high pressure pipe.		Spanner	
7. Prime several times the priming shaft priming handle.	Air comes out to the fuel injection pump side.	Priming handle	
8. Tighten the fuel high pressure pipe.		Spanner	
9. Perform priming again.	Fuel injection sound, "bitz" "bitz" can be heard.		
10. Perform air venting according to the order of 4~6.			

(5) Lapping of Valve Washer of Cylinder Head

If the suction and exhaust valve seats are being affected due to carbon deposit, or wear or corrosion (especially due to poor quality fuel oil), such may cause compression leakage. It is therefore necessary to remove the carbon and lap the valve seats.

1) **Lapping of Valve Seats**

Procedure	Instruction	Tool Used	Figure
1. Disassemble the cylinder head, and remove carbon from the valve seats.	Take care not to scratch the seats.		
2. Lap the valve seats with lapping compounds.	1) Do not mistake setting the mark of valve and the valve seat. (Differentiate the suction valve and exhaust valve.) 2) Apply the lapping compound to the valve seats and lap each other. Apply first coarse lapping compound, then fine one. Finish up the lapping with lubricating oil. Take care not to permit lapping compounds to enter into the vlave guides.	Lapping tool Lapping compounds (fine & coarse)	
3. After completion of lapping check the contact.	After wipe off the lapping compounds, sparingly apply blue paint or red lead on the valve seat, and check the contact. The contact is perfect if it is continuous.	Blue paint or red lead	
4. Clean the valve and valve seat.	Remove lapping compounds off the valve and valve seat with clean oil.		

If the contact of valve seats is inproper due to wear, grind the valve seat by a valve seat grinder available in the market, then lap them.
(Angle of valve seat: 90°)
If the valve seat is worn considerably, it should be changed.
Proper contacting width of seat part is 1.5 ~ 3.0.

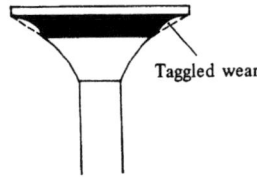
Taggled wear

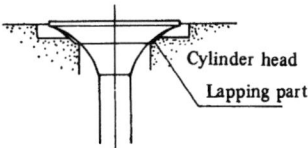
Cylinder head
Lapping part

7-3. Maintenance Standards of Main Parts

1. Limit for Wear

As the engine is used for a long time, each part will wear out reducing the performance or causing breaking or damage of crank shaft and piston pin unless such defective parts are replaced. Figures to be referred for replacing the parts are given in the following table. In the practical operation, however, no trouble will immediately occur when wear of some parts have advanced exceeding these figures.

Time should always be taken into consideration in judging wear of parts. Therefore, care must be taken to avoid customer's misunderstanding in using the figures in the table.

2. Maintenance Standard

Name of Parts					Nominal size	Difference in size	Standard Clearance of Ass'y	Maximun Allowable Clearance	Limit of Use	Measures to be taken
Cylinder liner	Inner dia.				105φ	+0.030 / 0			Up to wearing of chrome plating	Change liner
	Lug of rubber packing						One side 0.6 ~ 0.75			Be sure to change when changing liner
Clearance between piston and cylinder liner	Outer dia. of piston	Top				−0.675 / −0.705	0.17 ~ 0.23		−0.35	Change piston
		Skirt			105φ	−0.170 / −0.200				
	Inner dia. of liner					+0.030 / 0				
Compression clearance					2.45	±0.1	2.35 ~ 2.55			
Piston ring	Contact clearance (against standard dia.)	Inside of liner	Pressure ring				0.3 ~ 0.5		1.8	Change rings
			Oil scraping ring				0.3 ~ 0.5		1.8	Change rings
		Free condition	Pressure ring	No. 1			Approx. 13.33			
				No. 2,3			Approx. 13.75			
			Oil scraping ring				Approx. 13.75			
	Clearance between piston ring and ring groove	Pressure ring	Breadth of ring	No. 1	3.5	−0.01	0.02 ~ 0.055	0.2	−0.10	Change rings or piston
				No. 2,3	2.5	−0.03				
			Breadth of groove	No. 1	3.5	+0.025			+0.15	
				No. 2,3	2.5	+0.010				
		Oil scraping ring	Breadth of ring		4.5	−0.01 / −0.03	0.02 ~ 0.055		−0.10	Change rings of piston
			Breadth of groove			+0.025 / +0.010			+0.15	
Piston pin	Inner dia. of piston boss					0 / −0.011	Clearance (Tightening) −0.20 ~ 0.07			Change
	Outer dia. of piston pin				38φ				−0.07	Change
	Inner dia. of piston bush					+0.060 / +0.030	Clearance 0.03 ~ 0.073	0.15		Change

Name of Parts				Nominal size	Difference in size	Standard Clearance of Ass'y	Maximun Allowable Clearance	Limit of Use	Measures to be taken
Crank shaft	Crank pin	Outer dia. of crank pin		70φ	−0.040 / −0.055	0.04 ~ 0.094	0.15	0.7	When uneven wear of pin reaches over 0.06mm, regrind the pin and use undersized metal. (size: −0.25, −0.5)
		Inner dia. of bearing							
	Journal	Dia.	Outer dia. of journal	70φ	−0.040 / −0.055	0.040 ~ 0.093	0.15		When uneven wear of pin reaches over 0.06mm, regrind the pin and use undersized metal. (size: −0.25, −0.5)
			Dia. of bearing		+0.038 / +0.018				
		Side clearance	1SM Breadth of journal	110	−0.054	0.15 ~ 0.30	0.3	0.5	When the metal is worn out beyond limit of use, use oversized metal 0.1, 0.2.
			1SM Standard brea of bearing						
			2, 3SM Breadth of journal	50	+0.025 / 0	0.09 ~ 0.14	0.3	0.5	When the metal is worn out beyond limit of use, use oversized metal 0.2, 0.4.
			2, 3SM Std. breadth of bearing						
Cam shaft	Dia.	Outer dia. of journal part		—	—	—	—	—	—
		Inner dia. of bearing		—	—	—	—	—	—
	Breadth	Standard breadth of journal		—	—	—	—	—	—
		Standard breadth of bearing		—	—	—	—	—	—
	Height of cam			39.5	±0.01		—	−0.5	Change
Suction & exh. valves	Clearance between valve rod & valve guide	Suc	Valve rod outer dia.	10φ	−0.043 / −0.060	0.043 ~ 0.075	0.25	−0.15	Change valve or valve guide
			Valve guide inner dia.		+0.015 / 0				
		Exh.	Valve rod outer dia.	10φ	−0.043 / −0.060	0.043 ~ 0.075	0.20	−0.15	
			Valve guide inner dia.		+0.015 / 0				
	Valve seat	Angle		90°	—	—	—	—	—
		Breadth		—	—	2.83	—	—	—
		Sink		0.5				1.5	Change valve seat (exhaust side)
	Top Clearance	Suction valve		—		0.15		0.10~0.20	—
		Exhaust valve		—		0.15		0.10~0.20	—

3. Exchange Standard for Wear of Main Parts

	Specification of Main Parts		Calculation Bases for Wear Limit	Standard Size (unit-mm)	Limit of Measument
Limit of Wear	Inner dia. of cylinder liner		0.003D	105φ	+0.3
	Outer dia. of piston skirt		0.0025d	105φ	−0.25
	Outer dia. of piston pin		0.005d	38φ	−0.19
	Inner dia. of piston pin metal		0.005D	38φ	+0.19
	Outer dia. of crankshaft pin part		0.003d	70φ	+0.20
	Outer dia. of crankshaft journal		0.003d	70φ	−0.20
	Inner dia. of crank pin metal		0.004D	70φ	+0.28
	Inner dia. of crank metal		0.004D	70φ	+0.28
	No. 1 piston ring	Breadth	——	3.5	−0.20
		Thickness	0.15t	4.5	−0.65
	No. 2 − 3 piston ring	Breadth	——	2.5	−0.20
		Thickness	0.15t	4.5	−0.65
	Oil ring	Breadth	——	4.5	−0.20
		Thickness	0.15t	4.5	−0.65
	Outer dia. cam shaft journal		0.003d	50φ 28φ	−0.15 −0.10
	Height of cam shaft cam part		——	41.5	−0.40
Clearance	Clearance between piston ring and piston groove		——	0.037	+0.40
	Clearance between piston pin and bush		0.0075d	0.050	+0.28
	Clearance between crank pin and crank pin metal		0.0055d	0.070	+0.42
	Clearance between crank journal and crank metal		0.0055d	0.070	+0.42

(4) Maintenance Standards of Main Parts

Measuring Item	Measuring Position	Remarks	Measuring Instrument
Inner dia. of Cylinder liner		Measure in (a)(b) directions at * marked position. (* mark is positioned at No. 1 piston ring of upper dead center.)	Cylinder gauge
Outer dia. of piston skirt		Measure in (a)(b) direction at * marked position of piston skirt.	Micrometer
Clearance between cylinder liner and piston		Insert piston skirt portion into the upper part of cylinder liner (No. 1 piston ring position of top dead center) and measure clearance in the directions of crank shaft motion and at right angle to the crank shaft.	Thickness gauge
Outer dia. of piston pin		Maximum wear measured in (a)(b) directions at central * marked position.	Micrometer
Inner dia. of piston pin metal		Maximum wear dimension measureing metal inner dia. in (a)(b) directions.	Cylinder gauge
Clearance between piston and piston pin metal		Maximum clearance measured horizontal and vertical directions.	Thickness gauge
Outer dia. of crankshaft journal and crank metal		Maximum wear dimension measured at * mark in (a)(b) directions.	Micrometer

Measuring Item	Measuring Position	Remarks	Measuring Instrument
Inner dia. of crank metals (handle side and flywheel side)		Measure in horizontal and vertical directions at ∗ marked position.	Cylinder gauge
Thickness & breadth of piston rings	Breadth / Thickness		Micrometer
Breadth of ring groove			Use new piston ring as thickness gauge
Clearance between piston ring and ring groove		Maximum clearance between ring and groove	Thickness gauge
Simple method of measuring inner dia. of cylinder liner or thickness of piston ring	Use new one for eigher cylinder liner or piston ring. Cylinder liner / Piston ring	Measure piston ring fitting dimension. Calculation of wear in inner dia. direction. $$\frac{A - 0.3}{3.14} = \text{wear}$$ Here, if the cylinder liner is new, it is the doubled amount of wear in thickness of ring. If a new ring is used, it is the amount of wear in inner dia. of cylinder liner.	Thickness guage

Measuring Item	Measuring Position	Remarks	Measuring Instrument
Height of cam part of cam shaft	Height	Maximum height of cam part	Micrometer

8. COUNTERMEASURES TO ENGINE TORUBLES

Engine troubles must be detected at their early stage and defective parts be repaired in order to prevent the damages from further development. The symptoms of troubles must be quickly found through the eyes, ears, smelling, touching or the report of the operator. Once the place of engine trouble is found, it should be decided what kind of repair is required.
To do this properly, it is necessary to carry out logically correct checkings.
In order to readily find and repair the part of trouble, the operator is required to know each system and the function of each part completely. If a trouble is caused by the misoperation by operator, he must be instructed so that the same trouble will never occur again. Here the main troubles of engine, the causes and measures are to be mentioned.

8-1. Engine Troubles, Causes and Measures

(1) When engine does not start or it is difficult to start.

Causes

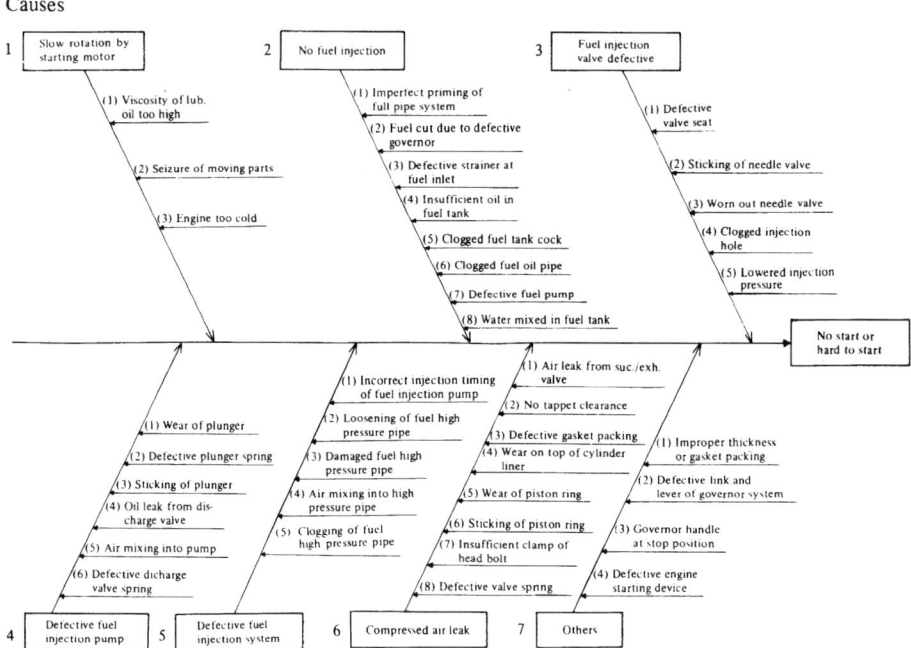

Measures

Causes	Measures	Causes	Measures
1–1	Warm up lub. oil or change	4–4	Fit the valve
1–2	Disassemble and adjust	4–5	Vent air
1–3	Warm up	4–6	Change
2–1	Carry out priming	5–1	Readjust
2–2	Readjust	5–2	Tighten firmly
2–3	Remove dust clogged	5–3	Change
2–4	Replenish fuel into tank	5–4	Vent air
2–5	Open the cock	6–5	Clean or replace with new one.
2–6	Clean	6–1	Fit the valve
2–7	Disassemble and adjust or change part	6–2	Readjust
2–8	Remove water through drain to empty the water in fuel pipe and carry priming	6–3	Change
		6–4	Change
		6–5	Change
3–1	Fit the part	6–6	Disassemble and adjust or change
3–2	Fit the part	6–7	Tighten clamp bolt evenly
3–3	Change	6–8	Change
3–4	Clean injection hole or change	7–1	Change
3–5	Readjust	7–2	Readjust
4–1	Change the set of plunger and barrel	7–3	Shift the governor handle at higher speed position
4–2	Change	7–4	Check and adjust
4–3	Disassemble and adjust or change		

(2) When power is insufficient

Causes

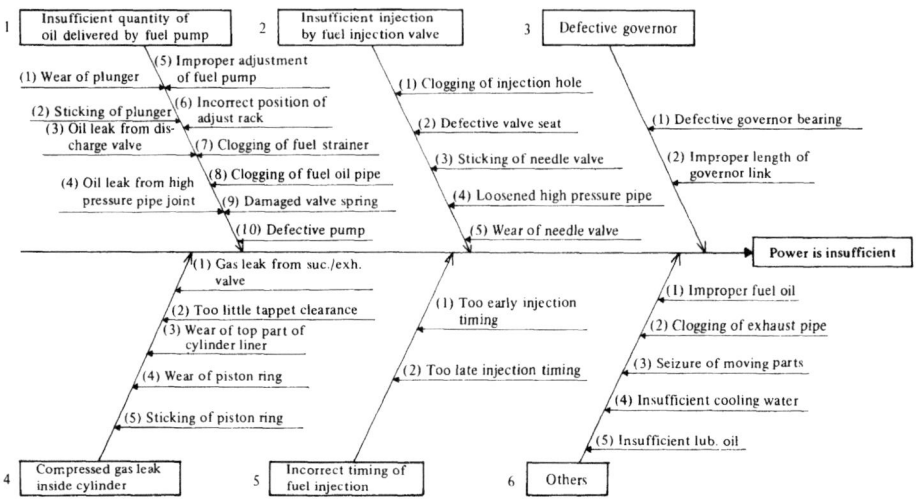

Measures

Causes	Measures	Causes	Measures
1−1	Change	4−1	Fit the valve
1−2	Disassemble and adjust or change	4−2	Readjust
		4−3	Change
1−3	Fit the valve	4−4	Change
1−4	Tighten firmly	4−5	Disassemble and adjust or change
1−5	Readjust		
1−6	Readjust		
1−7	Clean	5−1	Delay injection timing
1−8	Clean	5−2	Quicken the timing
1−9	Change	6−1	Change it with good oil
1−10	Repair	6−2	Clean
2−1	Clean injection hole or change	6−3	Disassemble and adjust
2−2	Fit the part or change	6−4	Fit the suction and discharge valves of cooling water pump
2−3	Fit the part or change		
2−4	Tighten firmly	6−5	Disassemble and clean lubricating oil pump and strainer
2−5	Change		
3−1	Change		
3−2	Repair		

(3) When revolution is not smooth

Causes

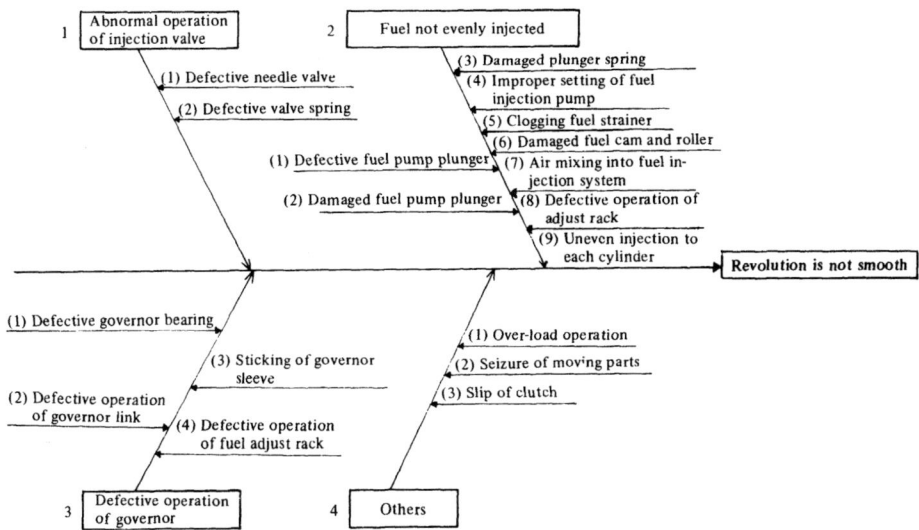

Measures

Causes	Measures	Causes	Measures
1–1	Fit the part	2–8	Repair
1–2	Change	2–9	Readjust
2–1	Clean	3–1	Change
2–2	Change	3–2	Repair
2–3	Change	3–3	Clean and adjust
2–4	Set pump correctly	3–4	Disassemble, clean and adjust
2–5	Clean	4–1	Reduce load
2–6	Change	4–2	Disassemble and adjust
2–7	Vent air and carry out priming	4–3	Check and repair

(4) When knocking takes place

Causes

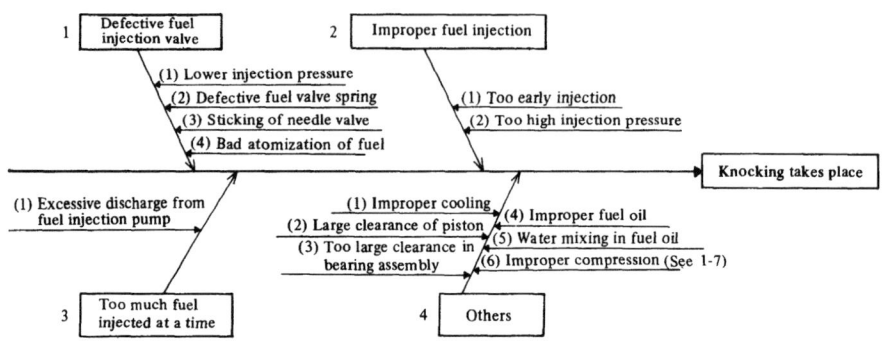

Measures

Causes	Measures	Causes	Measures
1−1	Adjust so that injection pressure becomes higher	3−1	Readjust the adjust rack of pump
1−2	Change	4−1	Check cooling water pump and fit the valve
1−3	Disassemble and fit	4−2	Change
1−4	Disassemble and adjust	4−3	Change
2−1	Delay injection timing	4−4	Change it with proper oil
2−2	Adjust it to specified injection pressure	4−5	Change it with proper oil
		4−6	Check and adjust

(5) When engine stops suddenly

Causes

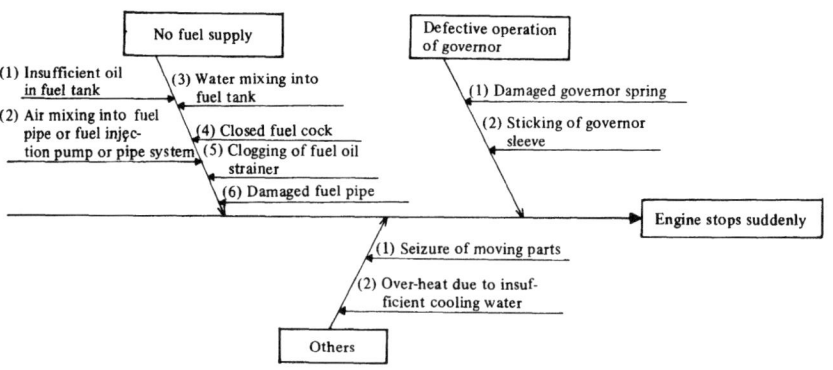

− 90 −

Measures

Causes	Measures	Causes	Measures
1—1	Supply fuel and carry out priming	2—1	Change
1—2	Vent air	2—2	Clean and adjust
1—3	Remove drain through drain hole, water in fuel pipe and carry out priming	3—1	Adjust or change
		3—2	Disassemble and adjust cooling water pump and check cooling water pipe
1—4	Check and repair, if necessary		
1—5	Clean		
1—6	Change		

(6) When powers of cylinders are not even

Causes

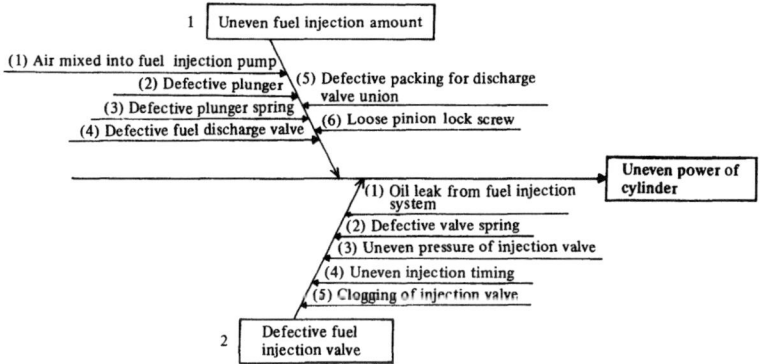

Measures

Causes	Measures	Causes	Measures
1—1	Vent air	2—1	Check and repair
1—2	Disassemble and clean	2—2	Change
1—3	Change	2—3	Readjust
1—4	Repair or change	2—4	Readjust
1—5	Change	2—5	Clean
1—6	Tighten firmly		

(7) When color of exhaust gas is bad

Causes

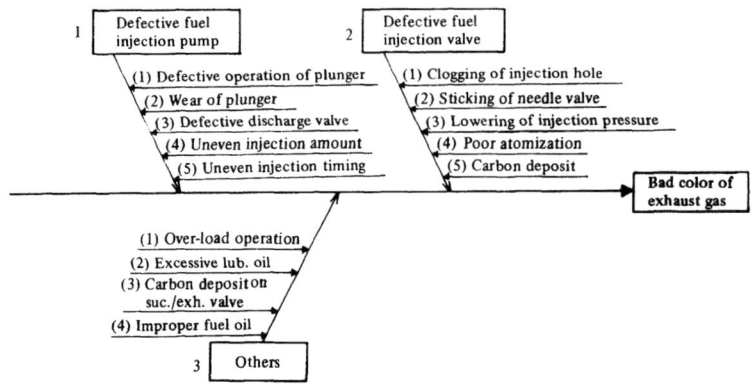

Measures

Causes	Measures	Causes	Measures
1–1	Check and repair or change	2–3	Readjust
1–2	Change	2–4	Repair or change
1–3	Check and repair or change	2–5	Clean
1–4	Readjust	3–1	Reduce load
1–5	Readjust	3–2	Adjust oil quantity
2–1	Clean	3–3	Clean
2–2	Repair or change	3–4	Change it with new fuel oil

(8) When other troubles take place

Causes

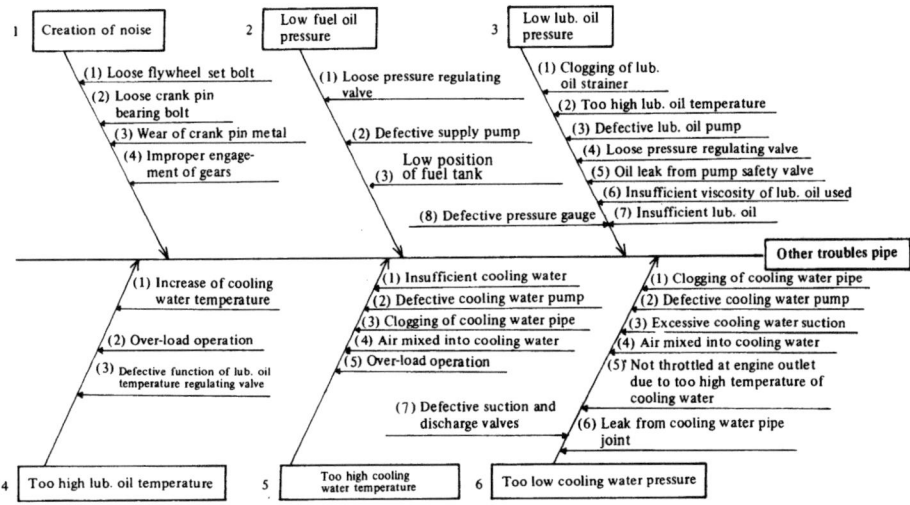

Measures

Causes	Measures	Causes	Measures
1−1	Tighten nut	4−1	Adjust return water or check and fit cooling water pump
1−2	Tighten nut and insert split pin	4−2	Decrease load
1−3	Remove adjust liner and adjust or change	4−3	Check and adjust
1−4	Check gears and change, if worn out	5−1	Widen throttled part
		5−2	Check and repair pump
		5−3	Clean
2−1	Tighten regulating valve	5−4	Check suction inlet
2−2	Check and repair	5−5	Decrease load
2−3	Rise piston	6−1	Clean
3−1	Clean	6−2	Check and repair pump
3−2	Increase cooling water	6−3	Drop suction or change pump
3−3	Disassemble and repair or change	6−4	Check suction hole
3−4	Tighten regulating valve	6−5	Remove causes increasing cooling water temp. and throttle
3−5	Tighten safety valve		
3−6	Change lub. oil	6−6	Check and repair
3−7	Supply oil	6−7	Check and repair
3−8	Change		

8-2. Causes and Measures for Reduction Reversing Gear

Causes

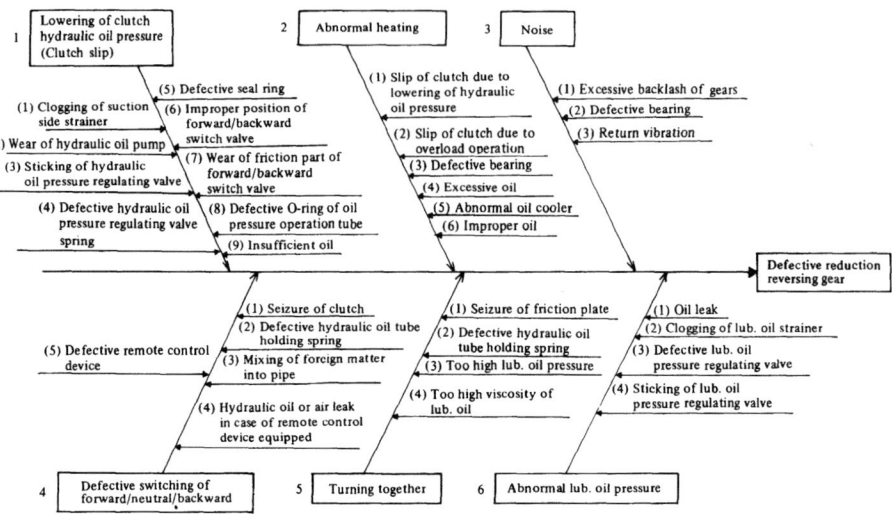

Measures

Causes	Measures	Causes	Measures
1−1	Disassemble and clean	3−1	Change
1−2	Adjust or change	3−2	Change
1−3	Adjust or change	3−3	Avoid revolution with vibration
1−4	Change	4−1	Change
1−5	Change	4−2	Change
1−6	Adjust	4−3	Clean
1−7	Change	4−4	Supply oil, adjust and change bellowphragm, etc.
1−8	Change		
1−9	Check oil leak and feed oil to specification	4−5	Adjust link
		5−1	Change
2−1	Check for (1) ~ (9) of 1	5−2	Change
2−2	Reduce load	5−3	Adjust lub. oil pressure valve
2−3	Change	5−4	Change oil with new one
2−4	Check oil quantity and adjust to specification	6−1	Check and adjust
		6−2	Disassemble and clean
2−5	Adjust minutely	6−3	Change
2−6	Change oil with new one	6−4	Adjust or change

9. STORAGE OF ENGINE

If the engine is not used for a long time, pay attention to the following storing instructions:

(1) In order to protect the engine from rust,
   i) Exchange lubricating oil and then carry out turning.
   ii) Supply oil to the link of fuel pump.
   iii) In case of electrical starting model, supply oil to the pinion shaft of cell motor.

 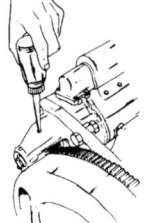

(2) To prevent rusting of the external part of engine, wipe off dirt and stain off the machine and apply a coat of oil over the external surface.
(3) To prevent the crank metal, piston, cylinder liner, etc. from rusting, crank the engine manually for about ten turns once a month.
(4) In case of electrical starting model, pay attention to storage of battery. Remove battery from the engine, and store at a place of low humidity. Charge up the battery once a month.
(5) When the engine is used in cold districts, it is necessary to drain the cooling water of the engine (including cooling water pump).
   Note: *When opening the cooling water cock the water in the engine is discharged. By taking out the plug, the water in the cooling water pump is released.*
(6) If rainwater may enter into the machine from the exhaust pipe, cover it.
(7) Set the suction and exhaust valves of No. 1, No. 2 and No. 3 cylinders to closed state and then stop the flywheel.
   Note: *In order to obtain the closed state of suction and exhaust valves as for No. 1, No. 2 and No. 3 cylinders set the No. 1 cylinder to position passing through its compression upper dead point (T) by 90° (where the position of flywheel start handle is coincident with the pointer) and set the decompression handle to the operating state.*

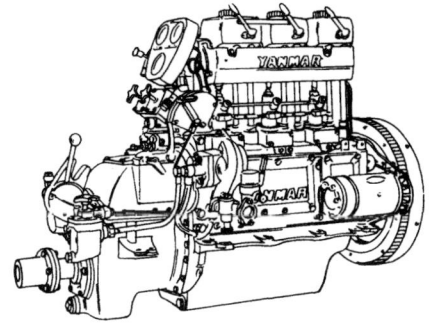

10. LIST OF APPROVED OILS

A) APPROVED DIESEL OILS

Supplier	Brand name
SHELL	Shell Diesoline or local equivalent
CALTEX	Caltex Diesel Oil
MOBIL	Mobile Diesel Oil
ESSO	Esso Diesel Oil

B) APPROVED LUBRICATING OILS

Supplier	Brand name	SAE NO.			
		Below 10°C	10°–20°C	20°–35°C	Over 35°
SHELL	Shell Rotella Oil	10W 20/20W	20/20W	30 40	50
	Shell Talona Oil	10W	20	30 40	50
	Shell Rimula Oil	20/20W	20/20W	30 40	
CALTEX	RPM Delo Marine Oil	10W	20	30 40	50
	RPM Delo Multi-Service-Oil	20/20W 10W	20	30 40	50
MOBIL	Delvac Special	10W	20	30 40	
	Delvac 20W–40	20W–40	20W–40		
	Delvac 1100 Series	10W 20–20W	20–20W	30 40	50
	Delvac 1200 Series	10W 20–20W	20–20W	30 40	50
ESSO	Estor HD	10W	20	30 40	
	Esso Lube HD		20	30 40	50
	Standard Diesel Oil	10W	20	30 40	50